SpringerBriefs in Probability and Mathematical Statistics

More information about this series at http://www.springer.com/series/14353

Dmitrii Silvestrov • Sergei Silvestrov

Nonlinearly Perturbed Semi-Markov Processes

Springer

Dmitrii Silvestrov
Department of Mathematics
Stockholm University
Stockholm, Sweden

Sergei Silvestrov
Division of Applied Mathematics
School of Education, Culture
and Communication (UKK)
Mälardalen University
Västerås, Sweden

ISSN 2365-4333 ISSN 2365-4341 (electronic)
SpringerBriefs in Probability and Mathematical Statistics
ISBN 978-3-319-60987-4 ISBN 978-3-319-60988-1 (eBook)
DOI 10.1007/978-3-319-60988-1

Library of Congress Control Number: 2017945357

Mathematics Subject Classification: 60J10, 60J22, 60J27, 60K15, 65C40

Printed on acid-free paper

This Springer imprint is published by Springer Nature
The registered company is Springer International Publishing AG
The registered company address is: Gewerbestrasse 11, 6330 Cham, Switzerland

Preface

The book presents new methods of asymptotic analysis for nonlinearly perturbed semi-Markov processes with finite phase spaces. These methods are based on special time-space screening procedures for sequential reduction of phase spaces for semi-Markov processes combined with the systematical use of operational calculus for Laurent asymptotic expansions.

We compose effective recurrent algorithms for the construction of Laurent asymptotic expansions, without and with explicit upper bounds for remainders, for power moments of hitting times for nonlinearly perturbed semi-Markov processes. We also illustrate the above results by getting asymptotic expansions for stationary and conditional quasi-stationary distributions of nonlinearly perturbed semi-Markov processes, in particular for birth-death-type semi-Markov processes, which play an important role in various applications.

It is worth noting that asymptotic expansions are a very effective instrument for studies of perturbed stochastic processes. The corresponding first terms in expansions give limiting values for properly normalized functionals of interest. The second terms let one estimate the sensitivity of models to small parameter perturbations. The subsequent terms in the corresponding expansions are usually neglected in standard linearization procedures used in studies of perturbed models. This, however, cannot be acceptable in the cases where values of perturbation parameter are not small enough. Asymptotic expansions let one take into account high-order terms in expansions, and in this way, to improve accuracy of the corresponding numerical procedures.

Semi-Markov processes are a natural generalization of discrete and continuous time Markov chains. These jump processes possess Markov property at moments of jumps and can have arbitrary distributions concentrated on a positive half-line for inter-jump times. In fact, this combination of basic properties makes semi-Markov processes a very flexible and effective tool for the description of queuing, reliability, and some biological systems, financial and insurance processes, and many other stochastic models.

As for Markov chains, a very important role in the theory of semi-Markov processes is played by hitting times and their moments. These random functionals are also known under such names as first-rare-event times, first passage times, and absorption times, in theoretical studies, and as lifetimes, failure times, extinction times, etc., in applications.

Expectations of hitting times play the key role in ergodic theorems and limit theorems of the law of large numbers type, due to dual relations connecting the corresponding stationary distributions and expectations of return times. Second moments appear in asymptotic results such as a central limit theorem. High-order moments appear in asymptotic results related to rates of convergence and asymptotical expansions in the above limit theorems as well as in large-deviation-type theorems. The moments of hitting times can also be effectively used for the estimation of tail probabilities for hitting times.

Models of perturbed Markov chains and semi-Markov processes, in particular for the most difficult cases of perturbed processes with absorption and so-called singularly perturbed processes, attracted attention of researchers in the middle of the twentieth century. An interest to these models has been stimulated by applications to control and queuing systems, information networks, epidemic models and models of population dynamics and mathematical genetics. We give references to related publications in the methodological and bibliographical remarks included in the book.

Markov-type processes with singular perturbations appear as natural tools for mathematical analysis of multicomponent systems with weakly interacting components. Asymptotics of moments for hitting-time type functionals and stationary distributions for corresponding perturbed processes play an important role in studies of such systems.

The role of perturbation parameters can be played by small failure probabilities or intensities in queuing and reliability systems and small mutation, extinction, or migration probabilities or intensities in biological systems. Perturbation parameters can also appear as artificial regularization parameters for decomposed systems, for example, as so-called damping parameters in information networks, etc.

In many cases, transition characteristics of the corresponding perturbed semi-Markov processes, in particular transition probabilities (of embedded Markov chains), and moments of transition times are nonlinear functions of a perturbation parameter, which admit asymptotic expansions with respect to this parameter.

Such Taylor and Laurent asymptotic expansions, respectively, for transition probabilities (of embedded Markov chains) and moments of transition times for perturbed semi-Markov processes, play the role of initial perturbation conditions in our studies. Two variants of these expansions are considered, with remainders given in the standard form $o(\cdot)$ and with explicit upper bounds for remainders.

We consider models with non-singular and singular perturbations, where the phase space for embedded Markov chains of pre-limiting perturbed semi-Markov processes is one class of communicative states, while the phase space for the limiting embedded Markov chain can possess an arbitrary communicative structure, i.e., can consist of one or several closed classes of communicative states and, possibly, a class of transient states.

The corresponding computational algorithms presented in the book have a universal character. They can be applied to perturbed semi-Markov processes with an arbitrary asymptotic communicative structure of phase spaces and are computationally effective due to the recurrent character of computational procedures.

The book includes six chapters, an appendix, and a bibliography.

In the introductory Chapter 1, we present in an informal form main problems, methods, and algorithms developed in the book, describe contents of the book by chapters, and give additional information for potential readers. Chapter 2 presents a calculus of Laurent asymptotic expansions, which serves as a basic analytic tool for asymptotic perturbation analysis of nonlinearly perturbed semi-Markov processes. Chapter 3 plays the key role in the book. We present here new algorithms of sequential phase space reduction for the construction of Laurent asymptotic expansions, without and with explicit upper bounds for remainders, for moments of hitting times for perturbed semi-Markov processes. In Chapter 4, the above asymptotic results are applied for getting asymptotic expansions for stationary distributions and related functionals for nonlinearly perturbed semi-Markov processes. In Chapter 5, the results presented in Chapter 4 are illustrated by the corresponding asymptotic results for nonlinearly perturbed birth-death-type semi-Markov processes, which play an important role in applications. In Chapter 6, we present some numerical examples illustrating results of previous chapters and give a brief survey of applied perturbed queuing systems, stochastic networks, and stochastic models of biological nature. In Appendix A, we make some methodological and bibliographical remarks and comment on new results presented in the book.

We hope that the publication of this new book related to asymptotic problems for perturbed stochastic processes will be a useful contribution to the continuing intensive studies in this area.

In addition to its use for research and reference purposes, the book can also be used in special courses on the subject and as a complementary reading in general courses on stochastic processes. In this respect, it may be useful for specialists as well as doctoral and advanced undergraduate students.

We would also like to thank all our colleagues at the Department of Mathematics, Stockholm University, and the Division of Applied Mathematics, School of Education, Culture and Communication, Mälardalen University, for creating an inspiring research environment and friendly atmosphere which stimulated our work.

Stockholm, Sweden
Västerås, Sweden
April 2017

Dmitrii Silvestrov
Sergei Silvestrov

Contents

List of Symbols

ε	Perturbation parameter
$L, h_A, k_A, l_{ij}^{\pm}, m_{ij}^{\pm}[k], M_{ij}^{\pm}[k], \ldots$	Parameters of asymptotic expansions
$a_k, a_{ij}[l], b_{ij}[k,l], c_i[l], \ldots$	Coefficients of asymptotic expansions
$o(\varepsilon^k), o_A(\varepsilon^k), o_i(\varepsilon^k), \dot{o}(\varepsilon^k), \ldots$	Remainders of asymptotic expansions
$\delta_A, G_A, \varepsilon_A, \tilde{\delta}_A, \tilde{G}_A, \tilde{\varepsilon}_A \ldots$	Parameters of upper bounds for remainders of asymptotic expansions
$\eta_{\varepsilon,n}$	Embedded Markov chain
${}_r\eta_{\varepsilon,n}, {}_{\bar{r}_{i,n}}\eta_{\varepsilon,n}, \ldots$	Reduced embedded Markov chains
$\eta_\varepsilon(t)$	Semi-Markov process
${}_r\eta_\varepsilon(t), {}_{\bar{r}_{i,n}}\eta_\varepsilon(t), {}_{\langle k,r\rangle}\eta_\varepsilon(t), \ldots$	Reduced semi-Markov processes
$\nu_{\varepsilon,i}$	Hitting time for embedded Markov chain
${}_r\nu_{\varepsilon,i}, {}_{\bar{r}_{i,n}}\nu_{\varepsilon,i}$	Hitting times for reduced embedded Markov chains
$\tau_{\varepsilon,i}$	Hitting time for semi-Markov process
${}_r\tau_{\varepsilon,i}, {}_{\bar{r}_{i,n}}\tau_{\varepsilon,i}, {}_{\langle k,r\rangle}\tau_{\varepsilon,j}, \ldots$	Hitting times for reduced semi-Markov processes
A, **B**, $\mathbf{C}_d$, **D**, $\mathbf{D}'$, ...	Conditions
$Q_{\varepsilon,ij}(t)$	Transition probability of semi-Markov process
${}_rQ_{\varepsilon,ij}(t)$	Transition probability of reduced semi-Markov process
$p_{ij}(\varepsilon), p_\pm(\varepsilon), \ldots$	Transition probabilities of embedded Markov chains
${}_rp_{ij}(\varepsilon), {}_{\bar{r}_{i,n}}p_\pm(\varepsilon), {}_{\langle k,r\rangle}p_{ij}(\varepsilon), \ldots$	Transition probabilities of reduced embedded Markov chains

$e_{ij}(\varepsilon), e_{\pm}(\varepsilon), e_i(\varepsilon), e(\varepsilon), \ldots$	Expectations of transition times for semi-Markov processes
${}_re_{ij}(\varepsilon),\ {}_{\bar{r}_{i,n}}e_{ij}(\varepsilon),\ {}_{\langle k,r\rangle}e_{ij}(\varepsilon), \ldots$	Expectations of transition times for reduced semi-Markov processes
$e_{ij}(k,\varepsilon)$	Power moment of transition time for semi-Markov process
${}_re_{ij}(k,\varepsilon),\ {}_{\bar{r}_{i,n}}e_{ij}(k,\varepsilon), \ldots$	Power moments of transition times for reduced semi-Markov processes
$E_{ij}(\varepsilon)$	Expectation of hitting time for semi-Markov process
${}_rE_{ij}(\varepsilon),\ {}_{\bar{r}_{i,n}}E_{ij}(\varepsilon), \ldots$	Expectations of hitting times for reduced semi-Markov processes
$E_{ij}(k,\varepsilon)$	Power moment of hitting time for semi-Markov process
${}_rE_{ij}(k,\varepsilon),\ {}_{\bar{r}_{i,n}}E_{ij}(k,\varepsilon), \ldots$	Power moments of hitting times for reduced semi-Markov processes
$\rho_i(\varepsilon)$	Stationary probability of embedded Markov chain
$\pi_i(\varepsilon)$	Stationary probability of semi-Markov process
$\pi_{\mathbb{U},i}(\varepsilon)$	Conditional quasi-stationary probability

Chapter 1
Introduction

This book is devoted to the study of asymptotic expansions for moment of hitting times, stationary and conditional quasi-stationary distributions, and other functionals, for nonlinearly perturbed semi-Markov processes. The introduction intends to present in informal form the main problems, methods, and algorithms that constitute the content of the book.

We present a number of simple examples of perturbed stochastic systems and try to show the logic and ideas that make a basis for new methods and algorithms of asymptotic analysis for nonlinearly perturbed semi-Markov processes developed in the book, using these examples. Also, we describe contents of the book by chapters and give an additional information for potential readers of the book.

1.1 Examples

We begin from simple examples which let us outline the circle of problems and illustrate results presented in this book. Then, we present and comment the content of the book by chapters.

1.1.1 Linear and Nonlinear Perturbations and High-Order Asymptotic Expansions

Let us begin with a simple example, which will let us explain how high order expansions can appear and be useful in the frame of studies of perturbed stochastic systems.

Let us consider the simplest M/M queuing system that consists of one server with exponentially distributed working and repairing times. The failure and the repairing

D. Silvestrov, S. Silvestrov, *Nonlinearly Perturbed Semi-Markov Processes*,
SpringerBriefs in Probability and Mathematical Statistics,
DOI 10.1007/978-3-319-60988-1_1

intensities are λ and μ, respectively. All working and repairing periods are supposed to be independent.

This queuing system can be interpreted as a simple model of a communication system with one channel or a power system with one electric generator.

Let $\eta(t)$ be a random variable which take value 0 or 1 if the above system, respectively, is repairing or working at instant t. According to the assumptions made above, $\eta(t), t \geq 0$ is, in this case, a continuous time homogeneous Markov chain with the phase space $\mathbb{X} = \{0,1\}$, continuous from the right trajectories and the generator matrix,

$$\mathbf{Q} = \left\| \begin{array}{cc} -\mu & \mu \\ \lambda & -\lambda \end{array} \right\| \qquad \begin{array}{c} \Leftarrow \langle 0 \rangle \\ \Leftarrow \langle 1 \rangle \end{array} \tag{1.1}$$

Figure 1.1 shows the transition graph, the transition probabilities $p_{ij}, i, j \in \mathbb{X}$ of the corresponding discrete time embedded Markov chain, and the expectations of exponential transition (sojourn) times $e_i, i \in \mathbb{X}$, for the process $\eta(t)$,

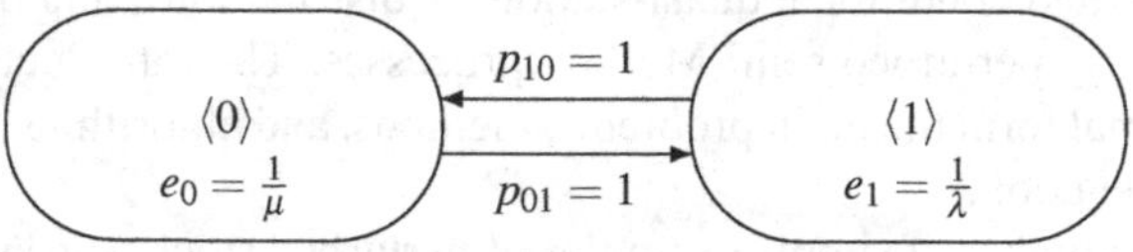

Fig. 1.1 M/M queuing system with one server. The transition graph for the Markov chain $\eta(t)$.

The standard object of interest is the stationary distribution of the Markov chain $\eta(t)$. As well known, the stationary probabilities are, in this case, the solution of well-known system of linear equations, $\pi_0\mu = \pi_1\lambda, \ \pi_0 + \pi_1 = 1$. They take the following form:

$$\pi_0 = \frac{\lambda}{\lambda + \mu}, \ \pi_1 = \frac{\mu}{\lambda + \mu}. \tag{1.2}$$

Let us assume that the above system has a very reliable server, i.e., the value of failure intensity λ is very small. An usual formalism here is to assume that intensity $\lambda(\varepsilon)$ depends on a small perturbation parameter $\varepsilon \in (0,1]$, in such a way that $\lambda(\varepsilon) \to 0$ as $\varepsilon \to 0$.

A study of questions related to actual meaning of parameter ε and the form and complexity of function $\lambda(\varepsilon)$ are beyond the frame of this book. It can be, just, the failure intensity itself. In this case, $\lambda(\varepsilon) = \varepsilon$. However, the role of parameter ε can be played, for example, by the reciprocal to cost of system, or by some of its physical parameter, or by some external environmental parameter.

We, just, make some general assumptions about the asymptotic behavior of function $\lambda(\varepsilon)$ as $\varepsilon \to 0$. In particular, the case, where $\lambda(\varepsilon) = \lambda_1\varepsilon + o(\varepsilon)$, i.e., this function is asymptotically linear as $\varepsilon \to 0$, can be interpreted as the case of linear type perturbation. Analogously, the case, where $\lambda(\varepsilon) = \lambda_1\varepsilon + \lambda_2\varepsilon^2 + o(\varepsilon^2)$, can be interpreted as the case of nonlinear quadratic type perturbation, the case, where $\lambda(\varepsilon) = \lambda_1\varepsilon + \lambda_2\varepsilon^2 + \cdots + \lambda_k\varepsilon^k + o(\varepsilon^k)$, as the case of nonlinear type perturbation of order k, etc. In this context, the case $\lambda(\varepsilon) = \varepsilon$ can, also, be considered as a particular case of nonlinear perturbation of order k (with coefficients $\lambda_2, \lambda_3, \ldots, \lambda_k = 0$ and remainder $o(\varepsilon^k) \equiv 0$), for any $k = 1, 2, \ldots$.

Here and henceforth, notations $o(\varepsilon^k), o_i(\varepsilon^k), \dot{o}(\varepsilon^k), \ldots$ are used for functions of parameter ε such that $\frac{o(\varepsilon^k)}{\varepsilon^k}, \frac{o_i(\varepsilon^k)}{\varepsilon^k}, \frac{\dot{o}(\varepsilon^k)}{\varepsilon^k}, \ldots \to 0$ as $\varepsilon \to 0$.

Let us consider the simplest case, where $\lambda(\varepsilon) = \varepsilon$. In this case, the corresponding stationary probabilities are nonlinear functions of ε, for example, $\pi_0(\varepsilon) = \frac{\varepsilon}{\varepsilon + \mu}$. The corresponding asymptotic expansions with one, two, and three terms for stationary probability $\pi_0(\varepsilon)$ have the forms, $\pi_0(\varepsilon) = \frac{1}{\mu}\varepsilon + o_1(\varepsilon)$, $\pi_0(\varepsilon) = \frac{1}{\mu}\varepsilon - \frac{1}{\mu^2}\varepsilon^2 + o_2(\varepsilon)$, and $\pi_0(\varepsilon) = \frac{1}{\mu}\varepsilon - \frac{1}{\mu^2}\varepsilon^2 + \frac{1}{\mu^3}\varepsilon^3 + o_3(\varepsilon)$. This relations lead to the use of approximations, $\pi_0^{(1)}(\varepsilon) = \frac{1}{\mu}\varepsilon$, $\pi_0^{(2)}(\varepsilon) = \frac{1}{\mu}\varepsilon - \frac{1}{\mu^2}\varepsilon^2$, and $\pi_0^{(3)}(\varepsilon) = \frac{1}{\mu}\varepsilon - \frac{1}{\mu^2}\varepsilon^2 + \frac{1}{\mu^3}\varepsilon^3$.

It is clear that, in the case, where parameter ε takes small but not very small values, the corrections terms $\frac{1}{\mu^2}\varepsilon^2$ and $\frac{1}{\mu^3}\varepsilon^3$ of two latter approximations may not be neglected.

For example, let $\mu = 0.5$ and $\varepsilon = 0.1$. In the case, the stationary probability $\pi_0(\varepsilon) \approx 0.166667$, the approximations $\pi_0^{(i)}(\varepsilon), i = 1,2,3$ take values, respectively, 0.2, 0.16, and 0.168. The corresponding errors, $|\pi_0(\varepsilon) - \pi_0^{(i)}(\varepsilon)|, i = 1,2,3$ take approximative values, respectively, 0.033333, 0.006667, and 0.001333. It is about, respectively, 20.0%, 4.0%, and 0.8% of $\pi_0(\varepsilon)$.

Analogously, in the case of nonlinear type perturbation, for example, of quadratic type, $\lambda(\varepsilon) = \lambda_1\varepsilon + \lambda_2\varepsilon^2 + o(\varepsilon^2)$, the standard procedure of linearization, which replaces the above expansion by the simpler linear one, $\lambda(\varepsilon) = \lambda_1\varepsilon + o(\varepsilon)$, can be not acceptable, if parameter ε takes values not small enough.

Of course, one can use the exact expressions for stationary probabilities for their asymptotic perturbation analysis, in the above example, since these stationary probabilities are solutions of the very simple linear system of equations with two unknowns. However, it can be a difficult problem in the case of multi-server or multi-component systems, where, in fact, stationary probabilities may be given as solutions of some system of linear equations of a high dimension and with nearly singular matrix of coefficients.

1.1.2 Laurent Asymptotic Expansions

In the second example, we try to clarify appearance of high order Laurent asymptotic expansions in problems related to asymptotic analysis of perturbed stochastic systems.

Let us consider a M/M queuing system that consists of two identical servers functioning independently of each other with exponentially distributed working and repairing times. The failure and the repairing intensities are λ and μ, respectively.

This queuing system can be interpreted as a simple model of a communication system with two channels or a power system with two electric generators.

Let $\eta(t)$ be the number of working servers at instant t. According to the assumptions made above, $\eta(t), t \geq 0$ is, in this case, a continuous time homogeneous Markov chain with the phase space $\mathbb{X} = \{0,1,2\}$, continuous from the right trajectories, and the generator matrix,

$$\mathbf{Q} = \left\| \begin{array}{ccc} -2\mu & 2\mu & 0 \\ \lambda & -(\lambda+\mu) & \mu \\ 0 & 2\lambda & -2\lambda \end{array} \right\| \begin{array}{c} \Leftarrow \langle 0 \rangle \\ \Leftarrow \langle 1 \rangle \\ \Leftarrow \langle 2 \rangle \end{array} \quad (1.3)$$

Let $0 = \zeta_0 < \zeta_1 < \cdots$ be instants of sequential jumps for the Markov chain $\eta(t)$, and $\eta_n = \eta(\zeta_n), n = 0, 1, \ldots$ be states of this Markov chain at sequential instants of jumps. As is known, η_n is a discrete time (embedded) Markov chain with the phase space $\mathbb{X}$ and the matrix of transition probabilities $\mathbf{P} = \|p_{ij}\|$, with elements given in Figure 1.2, which also shows the transition graph of the embedded Markov chain η_n, and the expectations $e_i, i = 1, 2, 3$ of exponential transition (sojourn) times for the process $\eta(t)$,

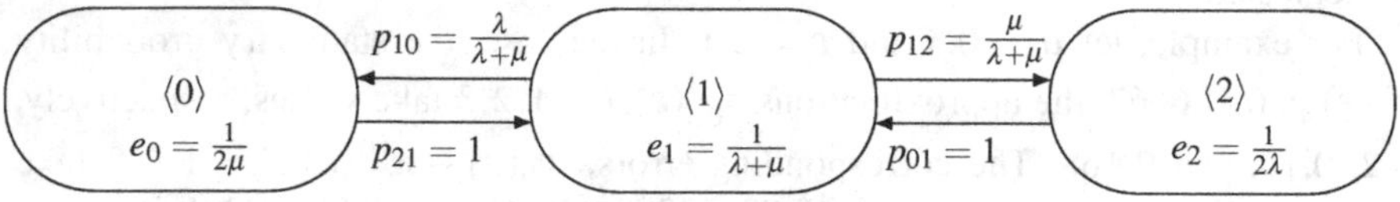

Fig. 1.2 M/M queuing system with two servers. The transition graph for the Markov chain $\eta(t)$.

Let us $\nu_j = \min(n \geq 1 : \eta_n = j)$ and $\tau_j = \zeta_{\nu_j}$ be the first hitting time to state $j \in \mathbb{X}$, respectively, for the embedded Markov chain η_n and for the continuous time Markov chain $\eta(t)$.

In this case, $E_{ij}(k) = \mathsf{E}_i \tau_j^k$ is the moment of order k for the first hitting time from state i to state j.

These moments are usually primary objects of interest in studies of such queuing systems. In particular, $E_{10}(1) = \mathsf{E}_1 \tau_0$ and $E_{20}(1) = \mathsf{E}_2 \tau_0$ are expectations for duration of one working period for the cases, where the number of working servers at the beginning of this working period is, respectively, 1 or 2. Also, $E_{01}(1)$ is the expectation for duration of one repairing period and $E_{00}(1)$ is the expectation of duration for one standard function cycle, which includes two sequential repairing and working periods. The high order moments $E_{ij}(k)$ can be useful in asymptotic results such as the central limit theorem, for finding upper bounds for tail probabilities of hitting times, etc.

Also, stationary probabilities $\pi_i, i = 1, 2, 3$ for the Markov chain $\eta(t)$ are, as usual, important objects of interest.

Expectations, $E_{ij}(1), i \in \mathbb{X}$ can be found, for every $j \in \mathbb{X}$, as solution of some well-known system of linear equations, which construction is based on analysis of an outcome of the first jump for the Markov chain $\eta(t)$. This system takes the form, $E_{ij}(1) = e_i + \sum_{r \neq j} p_{ir} E_{rj}(1), i \in \mathbb{X}$.

As far as the corresponding stationary probabilities are concerned, they can be found using the well-known system of linear equations $\pi_i = \sum_{j \neq i} \pi_j p_{ji}, i \in \mathbb{X}$, $\sum_{i \in \mathbb{X}} \pi_i = 1$, or using relations, which express stationary probabilities in the form of quotients $\pi_i = \frac{e_i}{E_{ii}(1)}, i \in \mathbb{X}$ of expectations of sojourn times e_i and expectations of return times $E_{ii}(1)$.

As in the first example, let us consider the simplest case, where the failure intensity $\lambda(\varepsilon) = \varepsilon$, for $\varepsilon \in (0, 1]$. In this case, the Markov chain $\eta(t) = \eta_\varepsilon(t)$, transition probabilities $p_{ij} = p_{ij}(\varepsilon)$, expectations of sojourn times $e_i = e_i(\varepsilon)$, as well as hitting times $\tau_j = \tau_{\varepsilon,j}$ and their moments $E_{ij}(k) = E_{ij}(k,\varepsilon)$, and stationary probabilities $\pi_i = \pi_i(\varepsilon)$ depend on parameter ε.

The corresponding simple computations yield that, for example, expectation $E_{00}(1,\varepsilon) = \frac{(\varepsilon+\mu)^2}{2\mu\varepsilon^2}$. Its asymptotic expansion with two terms has the form, $E_{00}(1,\varepsilon) = \frac{\mu}{2}\varepsilon^{-2} + \varepsilon^{-1} + \dot{o}_{00}(\varepsilon^{-1})$, i.e., it is a Laurent asymptotic expansion with the first term of the order ε^{-2}.

Let us note that expectation $E_{00}(1,\varepsilon)$ would be of the order ε^{-n} for an analogue of the above queuing system with n identical servers.

As far as stationary probabilities are concerned, they take values in interval $[0,1]$ and, thus, their asymptotic expansions can be only of Taylor type. For example, the stationary probability $\pi_0(\varepsilon) = \frac{\varepsilon^2}{(\varepsilon+\mu)^2}$ and its asymptotic expansion with two terms has the form,

$$\pi_0(\varepsilon) = \frac{1}{\mu^2}\varepsilon^2 - \frac{2}{\mu^3}\varepsilon^3 + o_0(\varepsilon^3). \tag{1.4}$$

The corresponding approximations can be obtained from the above asymptotic expansions by omitting their remainders. As in the first example, the second and other high order terms may be not neglected in these approximations, if the perturbation parameter ε takes values not small enough.

As in the first example, one can use the exact expressions for expectations of hitting times and stationary probabilities for their asymptotic perturbation analysis, since these expectations and probabilities are solutions of the simple linear systems of equations with three unknowns. However, it can be a difficult problem in the case of multi-server or multi-component systems, where, in fact, expectations of hitting times and stationary probabilities may be given as solutions of some systems of linear equations of high dimensions and with nearly singular matrices of coefficients. It would be desirable to develop some alternative computationally effective algorithms.

1.1.3 Recurrent Algorithms of Phase Space Reduction

Let us now try to compose some recurrent algorithms for construction of asymptotic expansions for expectations of hitting times and stationary probabilities, in the above example.

The Markov chain $\eta_\varepsilon(t)$ is a particular case of a semi-Markov process. In the semi-Markov setting, it is convenient to use the matrix of transition probabilities for the embedded Markov chain, $\|p_{ij}(\varepsilon)\|$, and the matrix of expectations of transition times, $\|e_{ij}(\varepsilon)\|$, which elements take, in this case, the form, $e_{ij}(\varepsilon) = e_i(\varepsilon)p_{ij}(\varepsilon), i,j \in \mathbb{X}$.

The matrix $\|p_{ij}(\varepsilon)\|$ and its representation in the form, where transition probabilities are given in the form of asymptotic expansions with two terms, have the following forms:

$$\left\|\begin{matrix} 0 & 1 & 0 \\ \frac{\varepsilon}{\varepsilon+\mu} & 0 & \frac{\mu}{\varepsilon+\mu} \\ 0 & 1 & 0 \end{matrix}\right\| \text{ and } \left\|\begin{matrix} 0 & 1 & 0 \\ 1-\frac{1}{\mu}\varepsilon+o_{10}(\varepsilon) & 0 & \frac{1}{\mu}\varepsilon-\frac{1}{\mu^2}\varepsilon^2+o_{12}(\varepsilon^2) \\ 0 & 1 & 0 \end{matrix}\right\|. \tag{1.5}$$

The matrix $\|e_{ij}(\varepsilon)\|$ and its representation in the form, where expectations are given in the form of asymptotic expansions with two terms, have the following forms:

$$\left\|\begin{matrix} 0 & \frac{1}{2\mu} & 0 \\ \frac{\varepsilon}{(\varepsilon+\mu)^2} & 0 & \frac{\mu}{(\varepsilon+\mu)^2} \\ 0 & \frac{1}{2\varepsilon} & 0 \end{matrix}\right\| \text{ and } \left\|\begin{matrix} 0 & \frac{1}{2\mu} & 0 \\ \frac{1}{\mu^2}\varepsilon-\frac{2}{\mu^3}\varepsilon^2+\dot{o}_{10}(\varepsilon^2) & 0 & \frac{1}{\mu}-\frac{2}{\mu^2}\varepsilon+\dot{o}_{12}(\varepsilon) \\ 0 & \frac{1}{2}\varepsilon^{-1} & 0 \end{matrix}\right\|. \tag{1.6}$$

Let us, for example, assume that the initial state $\eta_\varepsilon(0) = 0$ and construct a recurrent algorithm for computing expectation $E_{00}(\varepsilon)$, the stationary probability $\pi_0(\varepsilon)$, and asymptotic expansions for these quantities. This algorithm is based on the special time-space screening procedure of phase space reduction for the process $\eta_\varepsilon(t)$.

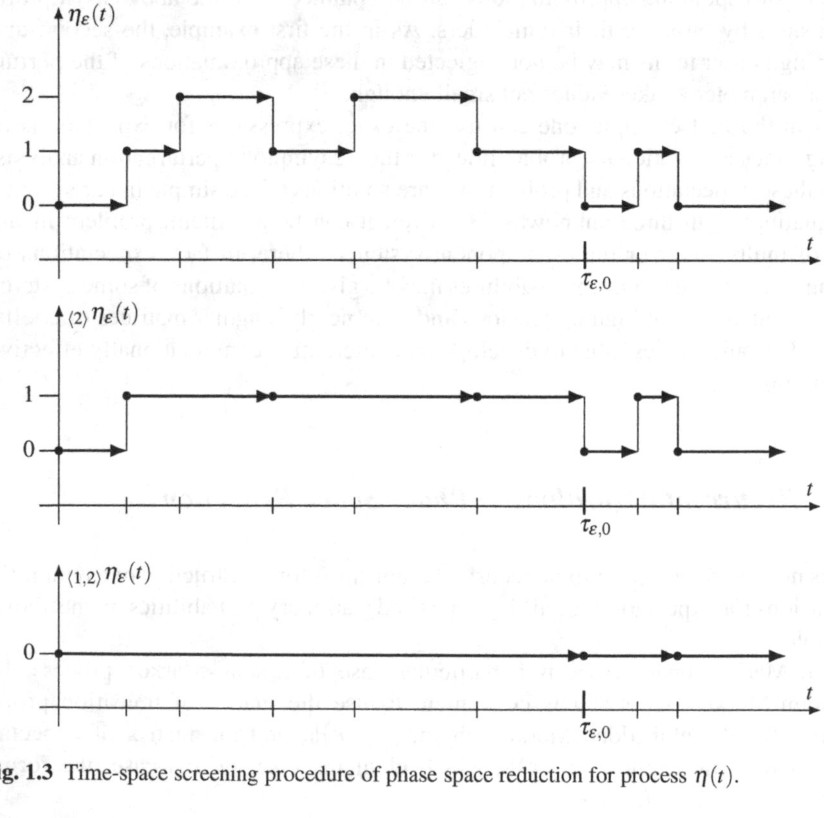

Fig. 1.3 Time-space screening procedure of phase space reduction for process $\eta(t)$.

Figure 1.3 illustrates transformation of trajectories for processes $\eta_\varepsilon(t)$ in the described above time-space screening procedure of phase space reduction.

Let us exclude state 2 from the phase $\mathbb{X} = \{0,1,2\}$ using the following time-space screening procedure. Let $0 = {}_{\langle 2\rangle}\zeta_{\varepsilon,0} < {}_{\langle 2\rangle}\zeta_{\varepsilon,1} < \cdots$ be sequential instants of hitting (resulted by jumps) of the reduced phase space ${}_{\langle 2\rangle}\mathbb{X} = \{0,1\}$ by the process $\eta_\varepsilon(t)$. Then, the reduced semi-Markov process can be defined as ${}_{\langle 2\rangle}\eta_\varepsilon(t) = \eta_\varepsilon({}_{\langle 2\rangle}\zeta_{\varepsilon,n})$, for ${}_{\langle 2\rangle}\zeta_{\varepsilon,n} \leq t < {}_{\langle 2\rangle}\zeta_{\varepsilon,n+1}, n = 0,1,\ldots$. A trajectory of this process is presented in the second part of Figure 1.3.

Instants ${}_{\langle 2\rangle}\zeta_{\varepsilon,n}, n = 1,2,\ldots$ are Markov moments for the Markov chain $\eta_\varepsilon(t)$ and, in sequel, for the reduced process ${}_{\langle 2\rangle}\eta_\varepsilon(t)$. However, inter-jump times ${}_{\langle 2\rangle}\kappa_{\varepsilon,n} = {}_{\langle 2\rangle}\zeta_{\varepsilon,n} - {}_{\langle 2\rangle}\zeta_{\varepsilon,n-1}, n = 0,1,\ldots$ have non-exponential distributions. That is why ${}_{\langle 2\rangle}\eta_\varepsilon(t), t \geq 0$ is a semi-Markov process but not a Markov chain.

Let us $\|{}_{\langle 2\rangle}p_{ij}(\varepsilon)\|$ and $\|{}_{\langle 2\rangle}e_{ij}(\varepsilon)\|$ be the corresponding 2×2 matrices of the transition probabilities and expectations of transition times, for the semi-Markov process ${}_{\langle 2\rangle}\eta_\varepsilon(t)$.

The above time-space screening procedure of phase space reduction can be repeated and applied to the process ${}_{\langle 2\rangle}\eta_\varepsilon(t)$. Let us now exclude state 1 from the phase space ${}_{\langle 2\rangle}\mathbb{X} = \{0,1\}$. Let $0 = {}_{\langle 1,2\rangle}\zeta_{\varepsilon,0} < {}_{\langle 1,2\rangle}\zeta_{\varepsilon,1} < \cdots$ be sequential instants of hitting (resulted by jumps) of the reduced phase space ${}_{\langle 1,2\rangle}\mathbb{X} = \{0\}$ by the process ${}_{\langle 2\rangle}\eta_\varepsilon(t)$. Then the reduced semi-Markov process can be defined as ${}_{\langle 1,2\rangle}\eta(t) = {}_{\langle 2\rangle}\eta_\varepsilon({}_{\langle 1,2\rangle}\zeta_{\varepsilon,n})$, for ${}_{\langle 1,2\rangle}\zeta_{\varepsilon,n} \leq t < {}_{\langle 1,2\rangle}\zeta_{\varepsilon,n+1}, n = 0,1,\ldots$. A trajectory of this process is presented in the third part of Figure 1.3.

Instants ${}_{\langle 1,2\rangle}\zeta_{\varepsilon,n}, n = 1,2,\ldots$ are Markov moments for the semi-Markov process ${}_{\langle 2\rangle}\eta_\varepsilon(t)$ and, in sequel, for the reduced process ${}_{\langle 1,2\rangle}\eta_\varepsilon(t)$. That is why the reduced process ${}_{\langle 1,2\rangle}\eta_\varepsilon(t), t \geq 0$ is also a semi-Markov process.

Let us, as above, $\|{}_{\langle 1,2\rangle}p_{00}(\varepsilon)\|$ and $\|{}_{\langle 1,2\rangle}p_{00}(\varepsilon)\|$ be 1×1 matrices of the transition probabilities and expectations of transition times for the semi-Markov process ${}_{\langle 1,2\rangle}\eta_\varepsilon(t)$. This process has the one-state phase space ${}_{\langle 1,2\rangle}\mathbb{X} = \{0\}$. The process ${}_{\langle 1,2\rangle}\eta_\varepsilon(t)$ occurs in state 0 after every jump. Thus, probability ${}_{\langle 1,2\rangle}p_{00}(\varepsilon) = 1$.

It is important that the above time-space screening procedure of phase space reduction preserves the return time $\tau_{\varepsilon,0}$. This time is the same for the processes $\eta_\varepsilon(t)$, ${}_{\langle 2\rangle}\eta_\varepsilon(t)$, and ${}_{\langle 1,2\rangle}\eta_\varepsilon(t)$.

Moreover, the return time $\tau_{\varepsilon,0}$ coincides with the moment of the first jump for the resulting one-state process ${}_{\langle 1,2\rangle}\eta(t)$, and, thus, the following very useful equality holds:

$$E_{00}(\varepsilon) = {}_{\langle 1,2\rangle}e_{00}(\varepsilon). \tag{1.7}$$

Elements of matrices $\|{}_{\langle 2\rangle}p_{ij}(\varepsilon)\|$ and $\|{}_{\langle 2\rangle}e_{ij}(\varepsilon)\|$ can be expressed as simple rational functions via elements of matrices $\|p_{ij}(\varepsilon)\|$ and $\|e_{ij}(\varepsilon)\|$, namely ${}_{\langle 2\rangle}p_{ij}(\varepsilon) = p_{ij}(\varepsilon) + p_{i2}(\varepsilon)\frac{p_{2j}(\varepsilon)}{1-p_{22}(\varepsilon)}$, $i,j \in {}_{\langle 2\rangle}\mathbb{X}$, and ${}_{\langle 2\rangle}e_{ij}(\varepsilon) = e_{ij}(\varepsilon) + e_{i2}(\varepsilon)\frac{p_{2j}(\varepsilon)}{1-p_{22}(\varepsilon)} + e_{22}(\varepsilon)\frac{p_{i2}(\varepsilon)}{1-p_{22}(\varepsilon)}\frac{p_{2j}(\varepsilon)}{1-p_{22}(\varepsilon)} + e_{2j}(\varepsilon)\frac{p_{i2}(\varepsilon)}{1-p_{22}(\varepsilon)}$, $i,j \in {}_{\langle 2\rangle}\mathbb{X}$.

These formulas reflect the fact that a jump from state i to state j can occur either via the direct jump from i to j, or via the jumps, first, from i to state 2, then a random number (which has a geometrical distribution with parameter $1 - p_{22}(\varepsilon)$) of jumps from 2 to 2, and finally the jump from state 2 to state j. The above formulas are written down in a general form, i.e., for the case, where probabilities $p_{rr}(\varepsilon), r \in \mathbb{X}$ can take a value in interval $[0,1)$. In the above example the probabilities $p_{rr}(\varepsilon) = 0, r \in \mathbb{X}$ that, in fact, simplify the above formulas.

Elements of matrices $\|_{\langle 1,2\rangle}p_{00}(\varepsilon)\|$ and $\|_{\langle 1,2\rangle}e_{00}(\varepsilon)\|$ can be expressed as analogous rational functions via elements of matrices $\|_{\langle 2\rangle}p_{ij}(\varepsilon)\|$ and $\|_{\langle 2\rangle}e_{ij}(\varepsilon)\|$, namely $_{\langle 1,2\rangle}p_{00}(\varepsilon) = {}_{\langle 2\rangle}p_{00}(\varepsilon) + {}_{\langle 2\rangle}p_{01}(\varepsilon)\frac{_{\langle 2\rangle}p_{10}(\varepsilon)}{1-{}_{\langle 2\rangle}p_{11}(\varepsilon)} = 1$, and $_{\langle 1,2\rangle}e_{00}(\varepsilon) = {}_{\langle 2\rangle}e_{00}(\varepsilon) + {}_{\langle 2\rangle}e_{01}(\varepsilon)\frac{_{\langle 2\rangle}p_{10}(\varepsilon)}{1-{}_{\langle 2\rangle}p_{11}(\varepsilon)} + {}_{\langle 2\rangle}e_{11}(\varepsilon)\frac{_{\langle 2\rangle}p_{01}(\varepsilon)}{1-{}_{\langle 2\rangle}p_{11}(\varepsilon)}\frac{_{\langle 2\rangle}p_{10}(\varepsilon)}{1-{}_{\langle 2\rangle}p_{11}(\varepsilon)} + {}_{\langle 2\rangle}e_{10}(\varepsilon)\frac{_{\langle 2\rangle}p_{01}(\varepsilon)}{1-{}_{\langle 2\rangle}p_{11}(\varepsilon)}$.

The above formulas have a clear recurrent character and let one effectively compute the above matrices of transition probabilities and expectations of transition times for reduced semi-Markov processes. In the above example, the only two steps are required. However, it is clear that the above recurrent algorithm can be effectively applied to multi-state processes, with large numbers of the above recurrent steps.

Moreover, the above recurrent formulas let one effectively compute the corresponding Laurent asymptotic expansions for elements of matrices of transition probabilities and expectations of transition times for reduced semi-Markov processes, by applying to these formulas standard operational rues (summation, multiplication, division) for Laurent asymptotic expansions.

The corresponding 2×2 matrices $\|_{\langle 2\rangle}p_{ij}\|$ and $\|_{\langle 2\rangle}e_{ij}\|$, obtained at the first step of the above recurrent algorithm, take the following forms:

$$\left\| \begin{matrix} 0 & 1 \\ \frac{\varepsilon}{\varepsilon+\mu} & \frac{\mu}{\varepsilon+\mu} \end{matrix} \right\| \text{ and } \left\| \begin{matrix} 0 & 1 \\ 1 - \frac{1}{\mu}\varepsilon + {}_{\langle 2\rangle}o_{10}(\varepsilon) & \frac{1}{\mu}\varepsilon - \frac{1}{\mu^2}\varepsilon^2 + {}_{\langle 2\rangle}o_{11}(\varepsilon^2) \end{matrix} \right\|. \tag{1.8}$$

and

$$\left\| \begin{matrix} 0 & \frac{1}{2\mu} \\ \frac{\varepsilon}{(\varepsilon+\mu)^2} & \frac{\mu(3\varepsilon+\mu)}{2\varepsilon(\varepsilon+\mu)^2} \end{matrix} \right\| \text{ and } \left\| \begin{matrix} 0 & \frac{1}{2\mu} \\ \frac{1}{\mu^2}\varepsilon - \frac{2}{\mu^3}\varepsilon^2 + {}_{\langle 2\rangle}\dot{o}_{10}(\varepsilon^2) & \frac{1}{2}\varepsilon^{-1} + {}_{\langle 2\rangle}\dot{o}_{11}(1) \end{matrix} \right\|. \tag{1.9}$$

The corresponding 1×1 matrices $\|_{\langle 1,2\rangle}e_{00}(\varepsilon)\| = \|E_{00}(\varepsilon)\|$, obtained at the second step of the above recurrent algorithm, take the following forms:

$$\left\| \frac{(\varepsilon+\mu)^2}{2\mu\varepsilon^2} \right\| \text{ and } \left\| \frac{\mu}{2}\varepsilon^{-2} + \varepsilon^{-1} + {}_{\langle 1,2\rangle}\dot{o}_{00}(\varepsilon^{-1}) \right\|. \tag{1.10}$$

Finally, the asymptotic expansion for the stationary probability $\pi_0(\varepsilon)$ can be found with the use of the above-mentioned formula, connecting stationary probabilities and expectations of sojourn and return times, relation (1.7) and the division rule for Laurent asymptotic expansions. This yields the asymptotic expansion for $\pi_0(\varepsilon)$, which coincides with the expansion given by relation (1.4),

$$\pi_0(\varepsilon) = \frac{e_0(\varepsilon)}{E_{00}(\varepsilon)} = \frac{\frac{1}{2\mu}}{\frac{\mu}{2}\varepsilon^{-2} + \varepsilon^{-1} + {}_{\langle 1,2\rangle}\dot{o}_{00}(\varepsilon^{-1})} = \frac{1}{\mu^2}\varepsilon^2 - \frac{2}{\mu^3}\varepsilon^3 + o_0(\varepsilon^3). \tag{1.11}$$

The recurrent algorithm based on the above time-space screening procedure of phase space reduction is computationally effective and can be applied to the semi-Markov processes with large numbers of states and for singularly perturbed models with phase space asymptotically splitting in several closed communicative classes and possibly a class of transient space. An analogous algorithm can also be composed for power moments of high order for hitting times of nonlinearly perturbed semi-Markov processes.

1.1.4 Rates of Convergence and Explicit Upper Bounds for Remainders of Asymptotic Expansions

Let us return to the first example and try to show that high order asymptotic expansions with explicit upper bounds for remainders are, in fact, analogous of relations, which give explicit estimates of rate of convergence in the simplest convergence relations. This interpretation relates to both Taylor and Laurent asymptotic expansions.

The simplest asymptotic convergence relation for the stationary probability $\pi_1(\varepsilon) = \frac{\mu}{\varepsilon+\mu} \to 1$ as $\varepsilon \to 0$ can be improved by the asymptotic relation, $|\pi_1(\varepsilon) - 1| \leq G_1\varepsilon^{\delta_1}$, for $\varepsilon \in (0, \varepsilon_1]$, with some values for parameters $\varepsilon_1 \in (0,1], G_1, \delta_1 > 0$. This asymptotic relation gives an explicit estimate for the rate of convergence in the asymptotic relation $\pi_1(\varepsilon) \to 1$ as $\varepsilon \to 0$. This convergence relation can be also re-written in the form of the simplest asymptotic expansion, $\pi_1(\varepsilon) = 1 + o_1(1)$, where the remainder $o_1(\varepsilon^0) = o_1(1) \to 0$ as $\varepsilon \to 0$, while the above estimate for rate of convergence one can be re-written in the form $|o_1(1)| \leq G_1\varepsilon^{\delta_1}$, for $\varepsilon \in (0, \varepsilon_1]$, i.e., as the relation, which gives explicit upper bounds for the remainder $o_1(1)$, asymptotically uniform with respect to perturbation parameter ε. The well-known Lagrange estimates for remainders of Taylor asymptotic expansions give, in this case, the following explicit formulas $\varepsilon_1 = 1$, $G_1 = \sup_{0<\varepsilon\leq 1} |\pi_1'(\varepsilon)| = \sup_{0<\varepsilon\leq 1} \frac{\mu}{(\varepsilon+\mu)^2} = \frac{1}{\mu}$, and $\delta_1 = 1$.

Analogously, the asymptotic expansion of the first order $\pi_1(\varepsilon) = 1 - \frac{1}{\mu}\varepsilon + o_2(\varepsilon)$ can be improved by the inequality $|o_2(\varepsilon)| \leq G_2\varepsilon^{1+\delta_2}$, for $\varepsilon \in (0, \varepsilon_2]$, with some parameters $\varepsilon_2 \in (0,1], G_2, \delta_2 > 0$, i.e., as the relation, which gives explicit upper bounds for the remainder $o_2(\varepsilon)$, asymptotically uniform with respect to perturbation parameter ε. The above-mentioned Lagrange estimates give, in this case, the following explicit formulas $\varepsilon_2 = 1$, $G_2 = \sup_{0<\varepsilon\leq 1} \frac{|\pi_1''(\varepsilon)|}{2} = \sup_{0<\varepsilon\leq 1} \frac{\mu}{(\varepsilon+\mu)^3} = \frac{1}{\mu^2}$, and $\delta_2 = 1$.

Let us now consider the same M/M queuing system, but assume that the failure intensity has the form, $\lambda = \lambda(\varepsilon)$, where $\lambda(\varepsilon)$ is a positive function of a perturbation parameter $\varepsilon \in (0,1]$, and $\lambda(\varepsilon) \to 0$ as $\varepsilon \to 0$.

If function $\lambda(\varepsilon)$ admits the asymptotic expansion of the first order, $\lambda(\varepsilon) = \lambda_1 \varepsilon + o_\lambda(\varepsilon)$, where λ_1 is a positive constant, then the stationary probability $\pi_1(\varepsilon) = \frac{\mu}{\lambda(\varepsilon)+\mu}$ also admits the first order Taylor asymptotic expansion, $\pi_1(\varepsilon) = 1 + \alpha_1 \varepsilon + \tilde{o}_\lambda(\varepsilon)$, with coefficient $\alpha_1 = -\frac{\lambda_1}{\mu}$ and remainder $\tilde{o}_\lambda(\varepsilon) = -\frac{\lambda_1 \alpha_1 \varepsilon^2 + o_\lambda(\varepsilon) + \alpha_1 o_\lambda(\varepsilon)}{\lambda(\varepsilon)+\mu}$.

If the remainder in the above asymptotic expansion for the failure intensity $\lambda(\varepsilon)$ admits explicit upper bounds of the form, $|o_\lambda(\varepsilon)| \leq G_\lambda \varepsilon^{1+\delta_\lambda}$, for $\varepsilon \in (0, \varepsilon_\lambda]$, with some parameters $\varepsilon_\lambda \in (0,1], G_\lambda, \delta_\lambda > 0$, then analogous explicit upper bounds of similar power form can be obtained for the remainder $\tilde{o}(\varepsilon)$. Indeed, the simple computations yield, in this case, the inequality, $|\tilde{o}_\lambda(\varepsilon)| \leq \tilde{G}\varepsilon^{1+\tilde{\delta}}$, for $\varepsilon \in (0, \tilde{\varepsilon}]$, where $\tilde{\varepsilon} = \varepsilon_\lambda$, $\tilde{G} = \frac{2(G_\lambda \mu + \lambda_1^2 + G_\lambda \lambda_1)}{\mu^2}$, and $\tilde{\delta} = 1 \wedge \delta_\lambda$.

The above example is typical for rates of convergence results. Usually, such results are based on assumptions about rates of convergence for some initial perturbed characteristics and algorithms, which let one get similar estimates for rates of convergence for some "global" functionals. Results concerning explicit upper bounds for remainders in asymptotic expansions have the same sense as explicit estimates of rates of convergence. The difference is that such upper bounds are obtained for remainders of high order expansions.

In the case of perturbed Markov and semi-Markov models the role of initial perturbed characteristics is played by transition probabilities of embedded Markov chains and moments of transition times, while global functionals can be moments of hitting times, stationary and quasi-stationary probabilities, etc. Both asymptotic expansions, with remainder given in the standard form of $o(\cdot)$, and remainders with explicit upper bounds, can be used in the corresponding perturbation conditions imposed on initial characteristics and be objects of interest in the asymptotic analysis of global functionals mentioned above.

It is also worth to note that methods based on the recurrent phase space reduction, shortly presented above can be modified and applied for getting asymptotic expansions with explicit upper bounds for remainders. This would only require to have the operational rules (summation, multiplication, division, and some others) for Laurent asymptotic expansions with explicit upper bounds for remainders. Then, one should apply these rules to the recurrent relations connecting transition characteristics of reduced semi-Markov process and to the formula connecting stationary probabilities and expectations of sojourn and hitting times.

1.2 Contents of the Book

The above discussion outlines shortly the range of problems which are objects of study in the present book. The book is devoted to the study of asymptotic expansions for nonlinearly perturbed semi-Markov processes with finite phase spaces. We concentrate our attention on power moments of hitting times, stationary distributions, and related functionals for perturbed semi-Markov processes. The asymptotic expansions for such functionals are obtained in two forms, without and with explicit upper bounds for remainders, for processes with an arbitrary asymptotic communicative structure of phase space. The methods are based on recurrent algorithms based on time-space screening procedures of phase space reductions and application of the calculus of Laurent asymptotic expansions, without and with explicit upper bound for remainders. A special attention is paid to the important class of perturbed semi-Markov processes of birth-death type.

1.2.1 Chapter 2

In Chapter 2, we present some kind of "operational calculus" for Laurent asymptotic expansions, which serves as a basic analytic tool for construction of the recurrent algorithms of phase space reduction for nonlinearly perturbed semi-Markov processes given in Chapters 3–6.

In Lemmas 2.1–2.3, we give explicit formulas for computing parameters, coefficients, and remainders for asymptotic expansions obtained as results of multiplication by constant, summation, multiplication, and division operations with Laurent asymptotic expansions, with remainders given in the standard form $o(\varepsilon^k)$. In Lemma 2.4, we present formulas for computing parameters, coefficients, and remainders connected with identities representing basic algebraic properties (commutative, associative, and distributive) for the above-mentioned operations with Laurent asymptotic expansions.

In Lemmas 2.5–2.7, we give explicit formulas for computing parameters of explicit power-type upper bounds, i.e., upper bounds of the form, $|o(\varepsilon^k)| \leq G_k \varepsilon^{k+\delta_k}$, $\varepsilon \in (0, \varepsilon_k]$, for remainders of expansions obtained as results of multiplication by constant, summation, multiplication, and division operations with Laurent asymptotic expansions. In Lemma 2.8 we present formulas for computing parameters of explicit power-type upper bounds for remainders connected with identities representing basic algebraic properties for the above operations.

1.2.2 Chapter 3

Chapter 3 plays the central role in the book. Here, we present basic recurrent algorithms for construction of Laurent asymptotic expansions for power moments of hitting times for nonlinearly perturbed semi-Markov processes.

We introduce a model of perturbed semi-Markov processes $\eta_\varepsilon(t), t \geq 0$ depending on some small perturbation parameter $\varepsilon \in (0, \varepsilon_0]$, for some $\varepsilon_0 \in (0, 1]$, and formulate basic perturbation conditions given in the form of asymptotic expansions for its transition characteristics (transition probabilities of embedded Markov chains and power moments of transition times up to some order $d \geq 1$). We study the model of singularly perturbed semi-Markov processes, where the phase space $\mathbb{X}$ of the processes $\eta_\varepsilon(t)$ is one class of communicative states, which can asymptotically split into one or several closed classes of communicative states and, possibly, a class of transient states, as $\varepsilon \to 0$.

We also describe a special time-space screening procedure of one-state reduction of the phase space $\mathbb{X}$ for a semi-Markov process $\eta_\varepsilon(t)$. According to this procedure, the reduced semi-Markov process ${}_{\langle r \rangle}\eta_\varepsilon(t)$ with the reduced phase space ${}_r\mathbb{X} = \mathbb{X} \setminus \{r\}$ makes jumps at sequential instants ${}_{\langle r \rangle}\zeta_{\varepsilon,n}$ of hitting the reduced phase space ${}_r\mathbb{X}$ by the initial semi-Markov process $\eta_\varepsilon(t)$ (resulted by its jumps) and takes values $\eta_\varepsilon({}_{\langle r \rangle}\zeta_{\varepsilon,n})$ at time intervals $[{}_{\langle r \rangle}\zeta_{\varepsilon,n-1}, {}_{\langle r \rangle}\zeta_{\varepsilon,n}), n = 0, 1, \ldots$. We get explicit formulas for transition characteristics of reduced semi-Markov process ${}_{\langle r \rangle}\eta_\varepsilon(t)$ and analogues of these characteristics related to hitting times of the initial semi-Markov process $\eta_\varepsilon(t)$ from the excluded state r to states from the reduced phase space ${}_r\mathbb{X} = \mathbb{X} \setminus \{r\}$, in the form of simple rational functions of the corresponding transition characteristics for the initial semi-Markov process. The above algorithm of one-state reduction is, just, one step in the multi-step recurrent algorithm of phase space reduction described below.

In Theorem 3.1, we prove invariance of hitting times $\tau_{\varepsilon,j}$ to states $j \in {}_r\mathbb{X}$ and, thus, invariance of their power moments $E_{ij}(k,\varepsilon) = \mathsf{E}_i \tau^k_{\varepsilon,j}, i \in \mathbb{X}, j \in {}_r\mathbb{X}$, with respect to the above time-space screening procedure.

In Theorems 3.2 and 3.3, we prove that the reduced semi-Markov processes ${}_{\langle r \rangle}\eta_\varepsilon(t)$ satisfy the same type perturbation conditions as the parent semi-Markov processes $\eta_\varepsilon(t)$. We, also, describe algorithms for re-calculation parameters, coefficients, and remainders of asymptotic expansions, which appear in these conditions, in terms of the corresponding parameters, coefficients and remainders of the asymptotic expansions, which appear in the perturbation conditions for processes $\eta_\varepsilon(t)$. These algorithms are based on application of operational rules for Laurent asymptotic expansions, with remainders given in the standard form $o(\cdot)$ presented in Lemmas 2.1–2.4, to the above rational functions representing the transition characteristics of the reduced semi-Markov processes. Here, the recurrent formulas connecting power moments of hitting times for semi-Markov processes are also used.

In Theorem 3.4, we describe the recurrent algorithm for construction of Laurent asymptotic expansions, with remainders given in the standard form $o(\cdot)$, for power moments of hitting times $E_{ij}(k,\varepsilon), i, j \in \mathbb{X}, k = 1, \ldots, d$. This algorithm is based on recurrent repetition of the above one-step time-space screening procedure up to exclusion from the initial phase space of all states $r \neq i, j$, except two arbitrary states $i \neq j$, and, then, exclusion of state i. In this way, we get Laurent asymptotic expansions for the transition characteristics of the reduced semi-Markov processes with one-state phase space $\{j\}$ and the power moments of transition times from state i to state j. These transition characteristics and power moments coincide with power moments $E_{lj}(k,\varepsilon), l = i, j, k = 1, \ldots, d$ for the semi-Markov processes $\eta_\varepsilon(t)$, since the above-mentioned invariance property of hitting times holds for every one-state reduction step. Thus, the above-mentioned Laurent asymptotic expansions also yield the Laurent asymptotic expansions, with remainders given in the standard form $o(\cdot)$, for the power moments $E_{lj}(k,\varepsilon), l = i, j, k = 1, \ldots, d$. We also prove, in Theorem 3.4, that the resulting asymptotic expansions are invariant with respect to the order, in which the states $r \neq i, j$ are excluded from the initial phase space.

In Theorems 3.5–3.7, we present analogous algorithms for getting Laurent asymptotic expansions, with explicit power upper bounds for remainders, for power moments of hitting times $E_{ij}(k,\varepsilon), i,j \in \mathbb{X}, k = 1,\ldots,d$, for nonlinearly perturbed semi-Markov processes.

1.2.3 Chapter 4

In Chapter 4, we apply results of Chapter 3 for getting asymptotic expansions for stationary distributions of nonlinearly perturbed semi-Markov processes.

In Theorems 4.1–4.2 and 4.3–4.4 we get asymptotic expansions, respectively, with remainders given in the standard for $o(\cdot)$ and with explicit power upper bounds for remainders, for stationary probabilities $\pi_i(\varepsilon), i \in \mathbb{X}$. The corresponding algorithms are based on the well-known relation $\pi_i(\varepsilon) = \frac{e_i(\varepsilon)}{E_{ii}(1,\varepsilon)}, i \in \mathbb{X}$, which connects the stationary probabilities $\pi_i(\varepsilon)$ with the expectations of sojourn times $e_i(\varepsilon)$ and the expectations of return times $E_{jj}(1,\varepsilon)$ and the use of division operational rule for Laurent asymptotic expansions given in Lemmas 2.2 and 2.6.

In Theorems 4.5–4.8, we get analogous asymptotic expansions, without and with explicit power upper bounds for remainders, for conditional quasi-stationary distributions and stationary limits for additive functionals for nonlinearly perturbed semi-Markov processes.

1.2.4 Chapter 5

In Chapter 5, we perform a detailed asymptotic perturbation analysis for nonlinearly perturbed finite birth-death-type semi-Markov processes, which play an important role in many applications. The corresponding results are obtained by applications of results presented in Chapters 2–4.

These semi-Markov processes have the special linear structure of phase space $\mathbb{X} = \{0,\ldots,N\}$, which implies a very important property of birth-death-type semi-Markov processes with respect to the time-space screening procedure of phase space reduction described in Chapter 2. The reduction of end states preserves the birth-death type for reduced semi-Markov processes. This makes it possible to get much more explicit recurrent formulas for parameters and coefficients of the corresponding asymptotic expansions and parameters of upper bound for remainders of these asymptotic expansions.

We consider three variants of perturbations, without asymptotically absorbing states, with one asymptotically absorbing end state 0, and with two asymptotically absorbing end states 0 and N. In Theorems 5.1 and 5.2, we get asymptotic expansions, with remainders given in the standard form, respectively, for stationary and conditional quasi-stationary distributions. In Theorems 5.3 and 5.4, we get asymptotic expansions, with explicit upper bounds for remainder, for stationary and conditional quasi-stationary distributions.

1.2.5 Chapter 6

In Chapter 6, we present some numerical examples illustrating results given in Chapters 2–5, and give a short survey of perturbed stochastic systems.

First, we discuss conditions that should be imposed on coefficients and remainders of asymptotic expansions for some real-valued functions $p_{ij}(\varepsilon), i, j \in \mathbb{X}$ and $e_{ij}(\varepsilon), i, j \in \mathbb{X}$, so that these functions could play roles of, respectively, transition probabilities of embedded Markov chains and expectations of transition times, for nonlinearly perturbed semi-Markov processes.

Then, we present a numerical example, where the above functions are given by concrete asymptotic expansions and show how asymptotic expansions, without and with explicit upper bounds for remainders, are computed for stationary probabilities of the corresponding perturbed semi-Markov processes, with the use of recurrent algorithms of sequential phase space reduction presented in Chapters 3 and 4.

We also present in this chapter a survey of perturbed queuing systems, stochastic networks, stochastic systems of biological nature, and some others, for which the asymptotic perturbation analysis plays an important role in their studies.

1.2.6 Appendix A

The book also includes an appendix with brief methodological and bibliographical remarks. Here, we comment methods used in the asymptotic analysis of perturbed Markov type models and new results presented in the book and formulate some prospective problems for future research in the area.

1.3 Conclusion

In this book, we present new algorithms for construction of asymptotic expansions for nonlinearly perturbed semi-Markov processes with finite phase spaces.

The method proposed in the book can be considered as a stochastic analogue of the Gauss elimination method. It is based on the special time-space screening procedure of sequential exclusion of states from phase spaces of perturbed semi-Markov processes accompanied by the algorithms for re-calculation of asymptotic expansions appearing in perturbation conditions for semi-Markov processes with reduced phase spaces. The corresponding algorithms are based on some kind of operational calculus for Laurent asymptotic expansions with remainders given in two forms, without and with explicit upper bounds.

The asymptotic analysis for perturbed processes of Markov type and related problems is a subject of intensive studies during several decades. However, the development of the theory, in particular, its part connected with asymptotic expansions, is still far from the completion. The book is concentrated in this area. The computational algorithms presented in the book have a universal character. They can be applied to perturbed semi-Markov processes with an arbitrary asymptotic communicative structure of phase spaces. The corresponding algorithms are computationally effective due to recurrent character of computational procedures.

The results presented in the book will be interesting to specialists, who work in such areas of the theory of stochastic processes as ergodic, limit, and large deviation theorems, analytical and computational methods for Markov chains, Markov, semi-Markov, and other classes of stochastic processes and their queuing, network, bio-stochastic, and other applications. We hope that the book will also attract attention of those researchers, who are interested in new analytical methods of analysis for nonlinearly perturbed stochastic processes and systems. It can also be useful for doctoral and advanced undergraduate students. This gives us hope that the book will find sufficient number of readers interested in stochastic processes and their applications.

Chapter 2
Laurent Asymptotic Expansions

In Chapter 2, we present operational rules for Laurent asymptotic expansions given in two forms, without and with explicit upper bounds for remainders. Here, we would like to refer to some classic books, for example, Hörmander (1990) and Markushevich (1985), where one can find basic facts about Laurent series, mainly, connected with representation problems for analytic functions. We, however, are interested in simpler objects and problems that are Laurent asymptotic expansions and operational rules for such expansions, with remainders given in the standard form $o(\cdot)$ and with explicit power-type upper bounds for remainders. In the former case, such rules should be possibly considered as generally known, while, in the latter case, asymptotic expansions have some novelty aspects.

2.1 Laurent Asymptotic Expansions with Remainders Given in the Standard Form

In Section 2.1, we present operational rules for Laurent asymptotic expansions with remainders given in the standard form.

2.1.1 Definition of Laurent Asymptotic Expansions with Remainders Given in the Standard Form

Let $A(\varepsilon)$ be a real-valued function defined on an interval $(0, \varepsilon_0]$, for some $0 < \varepsilon_0 \leq 1$, and given on this interval by a Laurent asymptotic expansion,

$$A(\varepsilon) = a_{h_A}\varepsilon^{h_A} + \cdots + a_{k_A}\varepsilon^{k_A} + o_A(\varepsilon^{k_A}), \tag{2.1}$$

where **(a)** $-\infty < h_A \leq k_A < \infty$ are integers, **(b)** coefficients $a_{h_A}, \ldots, a_{k_A}$ are real numbers, **(c)** function $o_A(\varepsilon^{k_A})/\varepsilon^{k_A} \to 0$ as $\varepsilon \to 0$.

D. Silvestrov, S. Silvestrov, *Nonlinearly Perturbed Semi-Markov Processes*,
SpringerBriefs in Probability and Mathematical Statistics,
DOI 10.1007/978-3-319-60988-1_2

We refer to such Laurent asymptotic expansion as a (h_A, k_A)-expansion with the remainder given in the standard form. This expansion is also a Taylor asymptotic expansion, if $h_A \geq 0$.

We say that (h_A, k_A)-expansion $A(\varepsilon)$ is pivotal if it is known that $a_{h_A} \neq 0$.

Parameter $L_A = k_A - h_A$ is called a length of the asymptotic expansion $A(\varepsilon)$.

Let a function $A(\varepsilon)$ be represented by a standard Taylor $(0, k_A)$-expansion. In applications, $A(\varepsilon)$ may be interpreted as a parameter for some perturbed system or process, and ε as a perturbation parameter. Respectively, $A(0) = a_0$ is interpreted as the value of above parameter for the corresponding unperturbed system or process. The asymptotic representation (2.1) reflects a stability property of parameter $A(\varepsilon)$ with respect to small perturbations, if $k_A = 0$. The asymptotic representation (2.1) corresponds to the model with linear type perturbation, if $k_A = 1$, and to the model with nonlinear type perturbation of order k_A if $k_A > 1$. The particular case of nonlinear type perturbation of order k_A is a polynomial type perturbation, where the corresponding remainder $o(\varepsilon^{k_A}) \equiv 0$. Also, an analytic-type perturbation, where function $A(\varepsilon)$ admits a representation in the form of absolutely convergent power series in interval $(0, \varepsilon_0]$, can be interpreted as a nonlinear type perturbation. Indeed, function $A(\varepsilon)$ can, in this case, be rewritten in the form (2.1), for any $k_A > 1$. In the general case, where $A(\varepsilon)$ is represented by a Laurent (h_A, k_A)-expansion, the perturbation type can be classified via the normalized function $\varepsilon^{-h_A} A(\varepsilon)$, which is a $(0, L_A)$-expansion. In this case, the perturbation type can be characterized as linear or nonlinear if, respectively, $L_A = 1$ or $L_A > 1$.

Let us now explain, why we can restrict consideration by the case, where parameter ε takes only positive values. Let us assume that function $A(\varepsilon)$ is also defined on some interval $[\varepsilon'_0, 0)$ and is given on this interval by a Laurent asymptotic expansion $A(\varepsilon) = A'(\varepsilon) = a'_{h'_A}\varepsilon^{h'_A} + \cdots + a'_{k'_A}\varepsilon^{k'_A} + o'_A(\varepsilon^{k'_A})$ analogous to the above one given in relation (2.1). Then $A'(\varepsilon), \varepsilon \in [\varepsilon'_0, 0)$ can always be rewritten as a function of positive parameter $-\varepsilon \in (0, -\varepsilon'_0]$ using the formula, $A'(\varepsilon) = A'(-(-\varepsilon)) = (-1)^{h'_A} a'_{h'_A} \cdot (-\varepsilon)^{h'_A} + \cdots + (-1)^{k'_A} a'_{k'_A} \cdot (-\varepsilon)^{k'_A} + o'_A((-1)^{k'_A}(-\varepsilon)^{k'_A})$. Thus, the operational analysis of function $A(\varepsilon)$, in particular computing of coefficients and estimation of remainder for the corresponding asymptotic expansion defined at a two-sided neighborhood of 0 can be reduced to analysis of two functions defined at positive one-sided neighborhoods of 0.

Let us also comment possible variations in representations of Laurent asymptotic expansions. If, for example, it is known that the (h_A, k_A)-expansion (2.1) has the first coefficient $a_{h_A} = 0$, then there is the obvious sense to exclude the first term $a_{h_A}\varepsilon^{h_A}$ from this asymptotic expansion and to rewrite it in the more informative form, $A(\varepsilon) = a_{h_A+1}\varepsilon^{h_A+1} + \cdots + a_{k_A}\varepsilon^{k_A} + o_A(\varepsilon^{k_A})$, i.e., as $(h_A + 1, k_A)$-expansion. Also, if it is known that the remainder of the (h_A, k_A)-expansion (2.1) can be represented in the form $o_A(\varepsilon^{k_A}) \equiv a_{k_A+1}\varepsilon^{k_A+1} + o'_A(\varepsilon^{k_A+1})$, where $o'_A(\varepsilon^{k_A+1})/\varepsilon^{k_A+1} \to 0$ as $\varepsilon \to 0$, then there is the obvious sense to replace the remainder $o_A(\varepsilon^{k_A})$ by the identical function $a_{k_A+1}\varepsilon^{k_A+1} + o'_A(\varepsilon^{k_A+1})$ in the asymptotic expansion (2.1) and rewrite it in the more informative form, $A(\varepsilon) = a_{h_A}\varepsilon^{h_A} + \cdots + a_{k_A}\varepsilon^{k_A} + a_{k_A+1}\varepsilon^{k_A+1} + o'_A(\varepsilon^{k_A+1})$, i.e., as $(h_A, k_A + 1)$-expansion.

Lemma 2.1. *If function* $A(\varepsilon) = a'_{h'_A}\varepsilon^{h'_A} + \cdots + a'_{k'_A}\varepsilon^{k'}_A + o'_A(\varepsilon^{k'_A}) = a''_{h''_A}\varepsilon^{h''_A} + \cdots + a''_{k''_A}\varepsilon^{k''_A} + o''_A(\varepsilon^{k''_A}), \varepsilon \in (0,\varepsilon_0]$ *can be represented as, respectively,* (h'_A, k'_A)*- and* (h''_A, k''_A)*-expansion, then the asymptotic expansion for function* $A(\varepsilon)$ *can be represented in the following most informative form* $A(\varepsilon) = a_{h_A}\varepsilon^{h_A} + \cdots + a_{k_A}\varepsilon^{k_A} + o_A(\varepsilon^{k_A}), \varepsilon \in (0,\varepsilon_0]$ *of* (h_A, k_A)*-expansion, with parameters* $h_A = h'_A \vee h''_A, k_A = k'_A \vee k''_A$*, and coefficients* $a_{h_A}, \ldots, a_{k_A}$*, and remainder* $o_A(\varepsilon^{k_A})$ *given by the following relations*:

(i) $a'_l = 0$*, for* $h'_A \leq l < h_A$ *and* $a''_l = 0$*, for* $h''_A \leq l < h_A$;

(ii) $a_l = a'_l = a''_l$*, for* $h_A \leq l \leq \tilde{k}_A = k'_A \wedge k''_A$;

(iii) $a_l = a''_l$*, for* $\tilde{k}_A = k'_A < l \leq k_A$ *if* $k'_A < k''_A$;

(iv) $a_l = a'_l$*, for* $\tilde{k}_A = k''_A < l \leq k_A$ *if* $k''_A < k'_A$;

(v) $\sum_{\tilde{k}_A < l \leq k'_A} a'_l\varepsilon^l + o'_A(\varepsilon^{k'_A}) = \sum_{\tilde{k}_A < l \leq k''_A} a''_l\varepsilon^l + o''_A(\varepsilon^{k''_A}), \varepsilon \in (0,\varepsilon_0]$ *and* $o_A(\varepsilon^{k_A})$ *coincides, for* $\varepsilon \in (0,\varepsilon_0]$*, with* $o''_A(\varepsilon^{k''_A})$ *if* $k'_A < k''_A$; $o'_A(\varepsilon^{k'_A}) = o''_A(\varepsilon^{k''_A})$ *if* $k'_A = k''_A$; *or* $o'_A(\varepsilon^{k'_A})$ *if* $k'_A > k''_A$.

The asymptotical expansion $A(\varepsilon)$ *is pivotal if and only if* $a_{h_A} = a'_{h_A} = a''_{h_A} \neq 0$.

Remark 2.1. A constant a can be interpreted as function $A(\varepsilon) \equiv a$. Thus, 0 can be represented, for any integer $-\infty < h \leq k < \infty$, as the (h,k)-expansion, $0 = 0\varepsilon^h + \ldots + 0\varepsilon^k + o(\varepsilon^k)$, with remainder $o(\varepsilon^k) \equiv 0$. Also, 1 can be represented, for any integer $0 \leq k < \infty$, as the $(0,k)$-expansion, $1 = 1 + 0\varepsilon + \ldots + 0\varepsilon^k + o(\varepsilon^k)$, with remainder $o(\varepsilon^k) \equiv 0$.

2.1.2 Operational Rules for Laurent Asymptotic Expansion with Remainders Given in the Standard Form

Let us consider four Laurent asymptotic expansions, $A(\varepsilon) = a_{h_A}\varepsilon^{h_A} + \cdots + a_{k_A}\varepsilon^{k_A} + o_A(\varepsilon^{k_A})$, $B(\varepsilon) = b_{h_B}\varepsilon^{h_B} + \cdots + b_{k_B}\varepsilon^{k_B} + o_B(\varepsilon^{k_B})$, $C(\varepsilon) = c_{h_C}\varepsilon^{h_C} + \cdots + c_{k_C}\varepsilon^{k_C} + o_C(\varepsilon^{k_C})$, and $D(\varepsilon) = d_{h_D}\varepsilon^{h_D} + \cdots + d_{k_D}\varepsilon^{k_D} + o_D(\varepsilon^{k_D})$ defined for $0 < \varepsilon \leq \varepsilon_0$, for some $0 < \varepsilon_0 \leq 1$.

Lemma 2.2. *The following operational rules take place for the above Laurent asymptotic expansions*:

(i) *If* $A(\varepsilon), \varepsilon \in (0,\varepsilon_0]$ *is a* (h_A, k_A)*-expansion and* c *is a constant, then* $C(\varepsilon) = cA(\varepsilon), \varepsilon \in (0,\varepsilon_0]$ *is a* (h_C, k_C)*-expansion such that*:

(a) $h_C = h_A, k_C = k_A$;

(b) $c_{h_C+r} = ca_{h_C+r}, r = 0, \ldots, k_C - h_C$;

(c) $o_C(\varepsilon^{k_C}) = co_A(\varepsilon^{k_A})$.

This expansion is pivotal if and only if $c_{h_C} = ca_{h_A} \neq 0$.

(ii) *If* $A(\varepsilon), \varepsilon \in (0,\varepsilon_0]$ *is a* (h_A, k_A)*-expansion and* $B(\varepsilon), \varepsilon \in (0,\varepsilon_0]$ *is a* (h_B, k_B)*-expansion, then* $C(\varepsilon) = A(\varepsilon) + B(\varepsilon), \varepsilon \in (0,\varepsilon_0]$ *is a* (h_C, k_C)*-expansion such that*:

(a) $h_C = h_A \wedge h_B$, $k_C = k_A \wedge k_B$;

(b) $c_{h_C+r} = a_{h_C+r} + b_{h_C+r}, r = 0, \ldots, k_C - h_C$*, where* $a_{h_C+r} = 0$*, for* $0 \leq r < h_A - h_C$ *and* $b_{h_C+r} = 0$*, for* $0 \leq r < h_B - h_C$;

(c) $o_C(\varepsilon^{k_C}) = \sum_{k_C < i \leq k_A} a_i \varepsilon^i + \sum_{k_C < j \leq k_B} b_i \varepsilon^j + o_A(\varepsilon^{k_A}) + o_B(\varepsilon^{k_B})$.

This expansion is pivotal if and only if $c_{h_C} = a_{h_C} + b_{h_C} \neq 0$.

(iii) *If* $A(\varepsilon), \varepsilon \in (0, \varepsilon_0]$ *is a* (h_A, k_A)*-expansion and* $B(\varepsilon), \varepsilon \in (0, \varepsilon_0]$ *is a* (h_B, k_B)*-expansion, then* $C(\varepsilon) = A(\varepsilon) \cdot B(\varepsilon), \varepsilon \in (0, \varepsilon_0]$ *is a* (h_C, k_C)*-expansion such that:*

(a) $h_C = h_A + h_B$, $k_C = (k_A + h_B) \wedge (k_B + h_A)$;

(b) $c_{h_C + r} = \sum_{0 \leq i \leq r} a_{h_A + i} b_{h_B + r - i}$, $r = 0, \ldots, k_C - h_C$;

(c) $o_C(\varepsilon^{k_C}) = \sum_{k_C < i+j, h_A \leq i \leq k_A, h_B \leq j \leq k_B} a_i b_j \varepsilon^{i+j} + \sum_{h_A \leq i \leq k_A} a_i \varepsilon^i o_B(\varepsilon^{k_B})$
$+ \sum_{h_B \leq j \leq k_B} b_j \varepsilon^j o_A(\varepsilon^{k_A}) + o_A(\varepsilon^{k_A}) o_B(\varepsilon^{k_B})$.

This expansion is pivotal if and only if $c_{h_C} = a_{h_A} b_{h_B} \neq 0$;

(iv) *If* $B(\varepsilon), \varepsilon \in (0, \varepsilon_0]$ *is a pivotal* (h_B, k_B)*-expansion, then there exists* $0 < \varepsilon_0' \leq \varepsilon_0$ *such that* $B(\varepsilon) \neq 0, \varepsilon \in (0, \varepsilon_0']$, *and* $C(\varepsilon) = \frac{1}{B(\varepsilon)}, \varepsilon \in (0, \varepsilon_0']$ *is a pivotal* (h_C, k_C)*-expansion such that:*

(a) $h_C = -h_B$, $k_C = k_B - 2h_B$;

(b) $c_{h_C} = b_{h_B}^{-1}$, $c_{h_C + r} = -b_{h_B}^{-1} \sum_{1 \leq i \leq r} b_{h_B + i} c_{h_C + r - i}$, $r = 1, \ldots, k_C - h_C$;

(c) $o_C(\varepsilon^{k_C}) = -\frac{\sum_{k_B - h_B < i+j, h_B \leq i \leq k_B, h_C \leq j \leq k_C} b_i c_j \varepsilon^{i+j} + \sum_{h_C \leq j \leq k_C} c_j \varepsilon^j o_B(\varepsilon^{k_B})}{b_{h_B} \varepsilon^{h_B} + \cdots + b_{k_B} \varepsilon^{k_B} + o_B(\varepsilon^{k_B})}$.

(v) *If* $A(\varepsilon), \varepsilon \in (0, \varepsilon_0]$ *is a* (h_A, k_A)*-expansion, and* $B(\varepsilon), \varepsilon \in (0, \varepsilon_0]$ *is a pivotal* (h_B, k_B)*-expansion, then there exists* $0 < \varepsilon_0' \leq \varepsilon_0$ *such that* $B(\varepsilon) \neq 0, \varepsilon \in (0, \varepsilon_0']$, *and* $D(\varepsilon) = \frac{A(\varepsilon)}{B(\varepsilon)}, \varepsilon \in (0, \varepsilon_0']$ *is a* (h_D, k_D)*-expansion such that:*

(a) $h_D = h_A + h_C = h_A - h_B$, $k_D = (k_A + h_C) \wedge (k_C + h_A)$
$= (k_A - h_B) \wedge (k_B - 2h_B + h_A)$;

(b) $d_{h_D + r} = \sum_{0 \leq i \leq r} a_{h_A + i} c_{h_C + r - i}$, $r = 0, \ldots, k_D - h_D$,

(c) $o_D(\varepsilon^{k_D}) = \sum_{k_D < i+j, h_B \leq i \leq k_B, h_C \leq j \leq k_C} b_i c_j \varepsilon^{i+j} + \sum_{h_B \leq i \leq k_B} b_i \varepsilon^i o_C(\varepsilon^{k_C})$
$+ \sum_{h_C \leq j \leq k_C} c_j \varepsilon^j o_B(\varepsilon^{k_B}) + o_B(\varepsilon^{k_B}) o_C(\varepsilon^{k_C})$,

where $c_{h_C + j}, j = 0, \ldots, k_C - h_C$ *and* $o_C(\varepsilon^{k_C})$ *are, respectively, the coefficients and the remainder of the* (h_C, k_C)*-expansion* $C(\varepsilon) = \frac{1}{B(\varepsilon)}$ *given in the above proposition* **(iv)**, *or by the following formulas:*

(d) $h_D = h_A - h_B$, $k_D = (k_A - h_B) \wedge (k_B - 2h_B + h_A)$;

(e) $d_{h_D + r} = b_{h_B}^{-1} (a_{h_A + r} - \sum_{1 \leq i \leq r} b_{h_B + i} d_{h_D + r - i})$, $r = 0, \ldots, k_D - h_D$;

(f) $o_D(\varepsilon^{k_D}) = \frac{\sum_{k_A \wedge (k_B + h_A - h_B) < l \leq k_A} a_l \varepsilon^l + o_A(\varepsilon^{k_A})}{b_{h_B} e^{h_B} + \cdots + b_{k_B} \varepsilon^{k_B} + o_B(\varepsilon^{k_B})}$

$$- \frac{\sum_{k_A \wedge (k_B + h_A - h_B) < i+j, h_B \leq i \leq k_B, h_D \leq j \leq k_D} b_i d_j \varepsilon^{i+j} + \sum_{h_D \leq j \leq k_D} d_j \varepsilon^j o_B(\varepsilon^{k_B})}{b_{h_B} \varepsilon^{h_B} + \cdots + b_{k_B} \varepsilon^{k_B} + o_B(\varepsilon^{k_B})}.$$

This expansion is pivotal if and only if $d_{h_D} = a_{h_A} c_{h_C} = a_{h_A} / b_{h_B} \neq 0$.

Remark 2.2. By Lemma 2.1, the Laurent asymptotic expansions for function $D(\varepsilon)$, given by the alternative formulas **(a)**–**(c)** and **(d)**–**(f)** in proposition **(v)** of Lemma 2.2, coincide. Also, these Laurent asymptotic expansions coincide with the expansions given by formulas **(a)**–**(c)** in propositions **(iv)** of Lemma 2.2, if $A(\varepsilon) \equiv 1$. In this case, 1 should be interpreted as the $(0, k_B - h_B)$-expansion, $1 = 1 + 0\varepsilon + \ldots + 0\varepsilon^{k_B - h_B} + o(\varepsilon^{k_B - h_B})$, with remainder $o(\varepsilon^{k_B - h_B}) \equiv 0$.

Let $A_m(\varepsilon) = a_{h_{A_m},m}\varepsilon^{h_{A_m}} + \cdots + a_{k_{A_m},m}\varepsilon^{k_{A_m}} + o_{A_m}(\varepsilon^{k_{A_m}}), \varepsilon \in (0, \varepsilon_0]$ be a (h_{A_m}, k_{A_m})-expansion, for $m = 1, \ldots, N$.

Lemma 2.3. *The following operational rules for multiple summation and multiplication take place for the above Laurent asymptotic expansions:*

(i) $B_n(\varepsilon) = A_1(\varepsilon) + \cdots + A_n(\varepsilon), \varepsilon \in (0, \varepsilon_0]$ *is, for every* $n = 1, \ldots, N$, *a* (h_{B_n}, k_{B_n})-*expansion, where:*

(a) $h_{B_n} = \min(h_{A_1}, \ldots, h_{A_n}),\ k_{B_n} = \min(k_{A_1}, \ldots, k_{A_n})$.

(b) $b_{h_{B_n}+l,n} = a_{h_{B_n}+l,1} + \cdots + a_{h_{B_n}+l,n},\ l = 0, \ldots, k_{B_n} - h_{B_n}$, *where*
$a_{h_{B_n}+l,m} = 0$, *for* $0 \leq l < h_{A_m} - h_{B_n}, m = 1, \ldots, n$.

(c) $o_{B_n}(\varepsilon^{k_{B_n}}) = \sum_{1 \leq m \leq n} \big(\sum_{k_{B_n} < l \leq k_{A_m}} a_{l,m}\varepsilon^l + o_{A_m}(\varepsilon^{k_{A_m}}) \big)$.

Expansion $B_n(\varepsilon)$ *is pivotal if and only if* $b_{h_{B_n},n} = a_{h_{A_1},1} + \cdots + a_{h_{A_n},n} \neq 0$.

(ii) $C_n(\varepsilon) = A_1(\varepsilon) \times \cdots \times A_n(\varepsilon), \varepsilon \in (0, \varepsilon_0]$ *is, for every* $n = 1, \ldots, N$, *a* (h_{C_n}, k_{C_n})-*expansion, where:*

(a) $h_{C_n} = h_{A_1} + \cdots + h_{A_n},\ k_{C_n} = \min(k_{A_l} + \sum_{1 \leq r \leq n, r \neq l} h_{A_r}, l = 1, \ldots, n)$.

(b) $c_{h_{C_n}+l,n} = \sum_{l_1+\cdots+l_n = l, 0 \leq l_m \leq k_{A_m} - h_{A_m}, 1 \leq m \leq n} \prod_{1 \leq m \leq n} a_{h_{A_m}+l_m,m}$,
$l = 0, \ldots, k_{C_n} - h_{C_n}$.

(c)
$$o_{C_n}(\varepsilon^{k_{C_n}}) = \sum_{k_{C_n} < l_1 + \cdots + l_n, h_{A_m} \leq l_m \leq k_{A_m}, 1 \leq m \leq n} \prod_{1 \leq m \leq n} a_{l_m,m}\varepsilon^{l_1 + \cdots + l_n}$$
$$+ \sum_{1 \leq r \leq n} \sum_{1 \leq m_1 < \cdots < m_r \leq n} \Big(\prod_{m \neq m_1, \ldots, m_r, 1 \leq m \leq n} \big(\sum_{h_{A_m} \leq l \leq k_{A_m}} a_{l,m}\varepsilon^l \big) \times \prod_{1 \leq i \leq r} o_{A_{m_i}}(\varepsilon^{k_{A_{m_i}}}) \Big).$$

Expansion $C_n(\varepsilon)$ *is pivotal if and only if* $c_{h_{C_n},n} = a_{h_{A_1},1} \times \cdots \times a_{h_{A_n},n} \neq 0$.

In particular, if $A_m(\varepsilon) = A_1(\varepsilon), \varepsilon \in (0, \varepsilon_0]$, *for* $m = 1, \ldots, N$, *then* $C_n(\varepsilon) = A_1(\varepsilon)^n$, $\varepsilon \in (0, \varepsilon_0]$ *is, for every* $n = 1, \ldots, N$, *a* (h_{C_n}, k_{C_n})-*expansion, where:*

(d) $h_{C_n} = nh_{A_1},\ k_{C_n} = k_{A_1} + (n-1)h_{A_1}$.

(e) $c_{h_{C_n}+l,n} = \sum_{l_1+\cdots+l_n = l, 0 \leq l_m \leq k_{A_1} - h_{A_1}, 1 \leq m \leq n} \prod_{1 \leq m \leq n} a_{h_{A_1}+l_m,1}$,
$l = 0, \ldots, k_{C_n} - h_{C_n}$.

(f)
$$o_{C_n}(\varepsilon^{k_{C_n}}) = \sum_{k_{C_n} < l_1 + \cdots + l_n, h_{A_m} \leq l_m \leq k_{A_m}, 1 \leq m \leq n} \prod_{1 \leq m \leq n} a_{l_m,1}\varepsilon^{l_1 + \cdots + l_n}$$
$$+ \sum_{1 \leq r \leq n} \binom{n}{r} \big(\sum_{h_{A_1} \leq l \leq k_{A_1}} a_{l,1}\varepsilon^l \big)^{n-r} \big(o_{A_1}(\varepsilon^{k_{A_1}}) \big)^r.$$

(iii) *Asymptotic expansions for functions* $B_n(\varepsilon) = A_1(\varepsilon) + \cdots + A_n(\varepsilon), n = 1, \ldots, N$ *and* $C_n(\varepsilon) = A_1(\varepsilon) \times \cdots \times A_n(\varepsilon), n = 1, \ldots, N$ *are invariant with respect to any permutation of summation and multiplication order in the above formulas.*

Remark 2.3. An alternative recurrent formulas for parameters, coefficients, and remainders of asymptotic expansions for sums $B_n(\varepsilon)$ and products $C_n(\varepsilon)$ can be obtained using the recurrent identities $B_n(\varepsilon) \equiv B_{n-1}(\varepsilon) + A_n(\varepsilon), n = 1, \ldots, N$, $B_0(\varepsilon) \equiv 0$, and $C_n(\varepsilon) \equiv C_{n-1}(\varepsilon) \cdot A_n(\varepsilon), n = 1, \ldots, N$, $C_0(\varepsilon) \equiv 1$ and applying to them the operational rules for Laurent asymptotic expansions given in Lemma 2.2.

2.1.3 *Algebraic Properties of Operational Rules for Laurent Asymptotic Expansions with Remainders Given in the Standard Form*

The following lemma summarizes some basic algebraic properties of Laurent asymptotic expansions. It is a corollary of Lemmas 2.1 and 2.2.

Lemma 2.4. *The summation and multiplication operations for Laurent asymptotic expansions defined in Lemma 2.2 possess the following algebraic properties, which should be understood as identities for the corresponding Laurent asymptotic expansions (i.e., identities for the corresponding parameters h, k, coefficients and remainders) of functions represented in two alternative forms in the functional identities given below:*

(i) *The summation and multiplication operations for Laurent asymptotic expansions satisfy the "elimination" identities that are implied by the corresponding functional identities,* $A(\varepsilon) + 0 \equiv A(\varepsilon)$, $A(\varepsilon) \cdot 1 \equiv A(\varepsilon)$, $A(\varepsilon) - A(\varepsilon) \equiv 0$, *and* $A(\varepsilon) \cdot A(\varepsilon)^{-1} \equiv 1$.

(ii) *The summation operation for Laurent asymptotic expansions is commutative and associative that is implied by the corresponding functional identities,* $A(\varepsilon) + B(\varepsilon) \equiv B(\varepsilon) + A(\varepsilon)$ *and* $(A(\varepsilon) + B(\varepsilon)) + C(\varepsilon) \equiv A(\varepsilon) + (B(\varepsilon) + C(\varepsilon))$.

(iii) *The multiplication operation for Laurent asymptotic expansions is commutative and associative that is implied by the corresponding functional identities,* $A(\varepsilon) \cdot B(\varepsilon) \equiv B(\varepsilon) \cdot A(\varepsilon)$ *and* $(A(\varepsilon) \cdot B(\varepsilon)) \cdot C(\varepsilon) \equiv A(\varepsilon) \cdot (B(\varepsilon) \cdot C(\varepsilon))$.

(iv) *The summation and multiplication operations for Laurent asymptotic expansions possess distributive property that is implied by the corresponding functional identity,* $(A(\varepsilon) + B(\varepsilon)) \cdot C(\varepsilon) \equiv A(\varepsilon) \cdot C(\varepsilon) + B(\varepsilon) \cdot C(\varepsilon)$.

Remark 2.4. In proposition **(i)** of Lemma 2.4, 0 should be interpreted as the (h_A, k_A)-expansion, $0 = 0 + 0\varepsilon^{h_A} + \ldots + 0\varepsilon^{k_A} + o(\varepsilon^{k_A})$, with remainder $o(\varepsilon^{k_A}) \equiv 0$, and 1 as $(0, k_A - h_A)$-expansion, $1 = 1 + 0\varepsilon + \ldots + 0\varepsilon^{k_A - h_A} + o(\varepsilon^{k_A - h_A})$, with remainder $o(\varepsilon^{k_A - h_A}) \equiv 0$.

Remark 2.5. The Laurent asymptotic expansion $A(\varepsilon)$ is assumed to be pivotal, in the elimination identity implied by functional identity $A(\varepsilon) \cdot A(\varepsilon)^{-1} \equiv 1$, and to hold, for $0 < \varepsilon \leq \varepsilon'_0$ such that $A(\varepsilon) \neq 0$, $\varepsilon \in (0, \varepsilon'_0]$.

2.2 Laurent Asymptotic Expansions with Explicit Upper Bounds for Remainders

In Section 2.2 we present operational rules for Laurent asymptotic expansions with explicit upper bounds for remainders.

2.2.1 *Definition of Laurent Asymptotic Expansions with Explicit Upper Bounds for Remainders*

Let $A(\varepsilon)$ be a real-valued function defined on an interval $(0, \varepsilon_0]$, for some $0 < \varepsilon_0 \leq 1$, and given on this interval by a Laurent asymptotic expansion,

$$A(\varepsilon) = a_{h_A}\varepsilon^{h_A} + \cdots + a_{k_A}\varepsilon^{k_A} + o_A(\varepsilon^{k_A}), \tag{2.2}$$

where **(a)** $-\infty < h_A \leq k_A < \infty$ are integers, **(b)** coefficients $a_{h_A}, \ldots, a_{k_A}$ are real numbers, **(c)** $|o_A(\varepsilon^{k_A})| \leq G_A \varepsilon^{k_A+\delta_A}$, for $0 < \varepsilon \leq \varepsilon_A$, where **(d)** $0 < \delta_A \leq 1, 0 \leq G_A < \infty$ and $0 < \varepsilon_A \leq \varepsilon_0$.

We refer to such Laurent asymptotic expansion as a $(h_A, k_A, \delta_A, G_A, \varepsilon_A)$-expansion.

The $(h_A, k_A, \delta_A, G_A, \varepsilon_A)$-expansion is also a (h_A, k_A)-expansion, according the definition given in Subsection 2.1.1, since, $o_A(\varepsilon^{k_A})/\varepsilon^{k_A} \to 0$ as $\varepsilon \to 0$.

We say that $(h_A, k_A, \delta_A, G_A, \varepsilon_A)$-expansion $A(\varepsilon)$ is pivotal if it is known that $a_{h_A} \neq 0$.

It is useful to note that there is no sense to consider, it seems, a more general case of upper bounds for the remainder $o_A(\varepsilon^{k_A})$, with parameter $\delta_A > 1$. Indeed, let us define $k'_A = k_A + [\delta_A] - \mathrm{I}(\delta_A = [\delta_A])$ and $\delta'_A = \delta_A - [\delta_A] + \mathrm{I}(\delta_A = [\delta_A]) \in (0, 1]$. The $(h_A, k_A, \delta_A, G_A, \varepsilon_A)$-expansion $A(\varepsilon)$ can be rewritten in the equivalent form of the $(h_A, k'_A, \delta'_A, G_A, \varepsilon_A)$-expansion, $A(\varepsilon) = a_{h_A}\varepsilon^{h_A} + \cdots + a_{k_A}\varepsilon^{k_A} + 0\varepsilon^{k_A+1} + \cdots + 0\varepsilon^{k'_A} + o'_A(\varepsilon^{k'_A})$, with the remainder $o'_A(\varepsilon^{k'_A}) = o_A(\varepsilon^{k_A})$, which satisfies inequalities $|o'_A(\varepsilon^{k'_A})| = |o_A(\varepsilon^{k_A})| \leq G_A \varepsilon^{k_A+\delta_A} = G_A \varepsilon^{k'_A+\delta'_A}$, for $0 < \varepsilon \leq \varepsilon_A$.

Let us also comment the analytic-type case, where function $A(\varepsilon)$ is given by an absolutely convergent power series, $A(\varepsilon) = \sum_{k \geq h_A} a_k \varepsilon^k$, in interval $(0, \varepsilon_0]$. In this case, $A(\varepsilon)$ can be represented in the form of $(h_A, k_A, 1, G_{k_A}, \varepsilon_{k_A})$-expansion $A(\varepsilon) = a_{h_A}\varepsilon^{h_A} + \cdots + a_{k_A}\varepsilon^{k_A} + o_{k_A}(\varepsilon^{k_A})$ with remainder $o_{k_A}(\varepsilon^{k_A}) = \sum_{k \geq k_A+1} a_k \varepsilon^k$, for any $k_A \geq h_A$. Indeed, remainder $o_{k_A}(\varepsilon^{k_A})$ satisfies, for any $\varepsilon_{k_A} \in (0, \varepsilon_0]$, inequalities, $|o_{k_A}(\varepsilon^{k_A})| \leq \varepsilon^{k_A+1} G_{k_A}, \varepsilon \in (0, \varepsilon_{k_A}]$, where $G_{k_A} = \sum_{k \geq k_A+1} |a_k| \varepsilon_{k_A}^{k-k_A-1}$.

The above remarks imply that the asymptotic expansion $A(\varepsilon)$ can be represented in different forms. In such cases, we consider forms with larger parameters h_A and k_A as more informative. As far as parameters δ_A, G_A, and ε_A are concerned, we consider as more informative forms, first, with larger values of parameter δ_A, second, with smaller values of parameter G_A and, third, with larger values of parameter ε_A.

Lemma 2.5. *If* $A(\varepsilon) = a'_{h'_A}\varepsilon^{h'_A} + \cdots + a'_{k'_A}\varepsilon^{k'_A} + o'_A(\varepsilon^{k'_A}) = a''_{h''_A}\varepsilon^{h''_A} + \cdots + a''_{k''_A}\varepsilon^{k''_A} + o''_A(\varepsilon^{k''_A}), \varepsilon \in (0, \varepsilon_0]$ *can be represented as, respectively,* $(h'_A, k'_A, \delta'_A, G'_A, \varepsilon'_A)$*- and* $(h''_A, k''_A, \delta''_A, G''_A, \varepsilon''_A)$*-expansion, then the* (h_A, k_A)*-expansion* $A(\varepsilon) = a_{h_A}\varepsilon^{h_A} + \cdots + a_{k_A}\varepsilon^k + o_A(\varepsilon^{k_A}), \varepsilon \in (0, \varepsilon_0]$ *given in Lemma 2.1 is an* $(h_A, k_A, \delta_A, G_A, \varepsilon_A)$*-expansion, with parameters* δ_A, G_A*, and* ε_A *chosen in the following way consistent with the priority order described above*:

$$(\delta_A, G_A, \varepsilon_A)$$

$$= \begin{cases} (\delta''_A, G''_A, \varepsilon''_A) & \text{if } k'_A < k''_A \text{ or } k'_A = k''_A, \delta'_A < \delta''_A, \\ (\delta'_A = \delta''_A, G'_A \wedge G''_A, \varepsilon'_A \wedge \varepsilon''_A) & \text{if } k'_A = k''_A, \delta'_A = \delta''_A, \\ (\delta'_A, G'_A, \varepsilon'_A) & \text{if } k'_A > k''_A \text{ or } k'_A = k''_A, \delta'_A > \delta''_A. \end{cases} \tag{2.3}$$

Remark 2.6. The following simple upper bounds take place for parameters G_A, δ_A, and ε_A given in Lemma 2.5, $\delta_A \geq \tilde{\delta}_A = \delta'_A \wedge \delta''_A$, $G_A \leq \tilde{G}_A = G'_A \vee G''_A$, and $\varepsilon_A \geq \tilde{\varepsilon}_A = \varepsilon'_A \wedge \varepsilon''_A$, which can be used in the case, where the priority order described above is ignored. Obviously, (h_A, k_A)-expansion $A(\varepsilon)$ is also a $(h_A, k_A, \tilde{\delta}_A, \tilde{G}_A, \tilde{\varepsilon}_A)$-expansion.

Remark 2.7. A constant a can be interpreted as function $A(\varepsilon) \equiv a$. Thus, 0 can be represented, for any integer $-\infty < h \leq k < \infty$, as the $(h, k, \delta_{h,k}, G_{h,k}, \varepsilon_{h,k})$-expansion, $0 = 0\varepsilon^h + \ldots + 0\varepsilon^k + o(\varepsilon^k)$, with remainder $o(\varepsilon^k) \equiv 0$ and, thus, parameters $\delta_{h,k} = 1$, $G_{h,k} = 0$, and $\varepsilon_{h,k} = \varepsilon_0$. Also, 1 can be represented, for any integer $0 \leq k < \infty$, as the $(0, k, \delta_k, G_k, \varepsilon_k)$-expansion, $1 = 1 + 0\varepsilon + \ldots + 0\varepsilon^k + o(\varepsilon^k)$, with remainder $o(\varepsilon^k) \equiv 0$ and, thus, parameters $\delta_k = 1$, $G_k = 0$, and $\varepsilon_k = \varepsilon_0$.

2.2.2 Operational Rules for Laurent Asymptotic Expansion with Explicit Upper Bounds for Remainders

Let us consider four Laurent asymptotic expansions, $A(\varepsilon) = a_{h_A}\varepsilon^{h_A} + \cdots + a_{k_A}\varepsilon^{k_A} + o_A(\varepsilon^{k_A})$, $B(\varepsilon) = b_{h_B}\varepsilon^{h_B} + \cdots + b_{k_B}\varepsilon^{k_B} + o_B(\varepsilon^{k_B})$, $C(\varepsilon) = c_{h_C}\varepsilon^{h_C} + \cdots + c_{k_C}\varepsilon^{k_C} + o_C(\varepsilon^{k_C})$, and $D(\varepsilon) = d_{h_D}\varepsilon^{h_D} + \cdots + d_{k_D}\varepsilon^{k_D} + o_D(\varepsilon^{k_D})$ defined for $0 < \varepsilon \leq \varepsilon_0$, for some $0 < \varepsilon_0 \leq 1$.

Let us denote, $F_A = \max_{h_A \leq i \leq k_A} |a_i|$, $F_B = \max_{h_B \leq i \leq k_B} |b_i|$, $F_C = \max_{h_C \leq i \leq k_C} |c_i|$, and $F_D = \max_{h_D \leq i \leq k_D} |d_i|$.

Lemma 2.6. *The above Laurent asymptotic expansions have the following operational rules for computing remainders:*

(i) *If $A(\varepsilon), \varepsilon \in (0, \varepsilon_0]$ is a $(h_A, k_A, \delta_A, G_A, \varepsilon_A)$-expansion and c is a constant, then $C(\varepsilon) = cA(\varepsilon), \varepsilon \in (0, \varepsilon_0]$ is a $(h_C, k_C, \delta_C, G_C, \varepsilon_C)$-expansion with parameters h_C, k_C and coefficients $c_r, r = h_C, \ldots, k_C$ given in proposition* **(i)** *of Lemma 2.2, and parameters δ_C, G_C, and ε_C given by the formulas:*

(a) $\delta_C = \delta_A = \tilde{\delta}_C$;

(b) $G_C = |c|G_A = \tilde{G}_C$;

(c) $\varepsilon_C = \varepsilon_A = \tilde{\varepsilon}_C = \tilde{\varepsilon}_C$.

(ii) *If $A(\varepsilon), \varepsilon \in (0, \varepsilon_0]$ is a $(h_A, k_A, \delta_A, G_A, \varepsilon_A)$-expansion and $B(\varepsilon), \varepsilon \in (0, \varepsilon_0]$ is a $(h_B, k_B, \delta_B, G_B, \varepsilon_B)$-expansion, then $C(\varepsilon) = A(\varepsilon) + B(\varepsilon)$, $\varepsilon \in (0, \varepsilon_0]$ is a $(h_C, k_C, \delta_C, G_C, \varepsilon_C)$-expansion with parameters h_C, k_C and coefficients $c_r, r = h_C, \ldots, k_C$ given in proposition* **(ii)** *of Lemma 2.2, and parameters δ_C, G_C, and ε_C given by formulas:*

(a) $\delta_C = \delta_A \mathrm{I}(k_A < k_B) + (\delta_A \wedge \delta_B)\mathrm{I}(k_A = k_B) + \delta_B \mathrm{I}(k_B < k_A) \geq \tilde{\delta}_C = \delta_A \wedge \delta_B$;

(b) $$G_C = \sum_{k_C < i \leq k_A} |a_i|\varepsilon_C^{i-k_C-\delta_C} + \sum_{k_C < j \leq k_B} |b_j|\varepsilon_C^{j-k_C-\delta_C} + G_A\varepsilon_C^{k_A+\delta_A-k_C-\delta_C} + G_B\varepsilon_C^{k_B+\delta_B-k_C-\delta_C}$$
$$\leq \tilde{G}_C = F_A(k_A - k_C) + G_A + F_B(k_B - k_C) + G_B;$$

(c) $\varepsilon_C = \varepsilon_A \wedge \varepsilon_B = \tilde{\varepsilon}_C$.

(iii) *If $A(\varepsilon), \varepsilon \in (0, \varepsilon_0]$ is a $(h_A, k_A, \delta_A, G_A, \varepsilon_A)$-expansion and $B(\varepsilon), \varepsilon \in (0, \varepsilon_0]$ is a $(h_B, k_B, \delta_B, G_B, \varepsilon_B)$-expansion, then $C(\varepsilon) = A(\varepsilon) \cdot B(\varepsilon)$, $\varepsilon \in (0, \varepsilon_0]$ is a $(h_C, k_C, \delta_C, G_C, \varepsilon_C)$-expansion with parameters h_C, k_C and coefficients $c_r, r = h_C, \ldots, k_C$ given in proposition* **(iii)** *of Lemma 2.2, and parameters δ_C, G_C, and ε_C given by formulas:*

(a) $\delta_C = \delta_A \mathrm{I}(k_A + h_B < k_B + h_A) + (\delta_A \wedge \delta_B)\mathrm{I}(k_A + h_B = k_B + h_A)$
$+ \delta_B \mathrm{I}(k_A + h_B > k_B + h_A) \geq \tilde{\delta}_C = \delta_A \wedge \delta_B$;

(b) $G_C = \sum_{k_C < i+j, h_A \leq i \leq k_A, h_B \leq j \leq k_B} |a_i||b_j|\varepsilon_C^{i+j-k_C-\delta_C}$
$+ G_A \sum_{h_B \leq j \leq k_B} |b_j|\varepsilon_C^{j+k_A+\delta_A-k_C-\delta_C} + G_B \sum_{h_A \leq i \leq k_A} |a_i|\varepsilon_C^{i+k_B+\delta_B-k_C-\delta_C}$
$+ G_A G_B \varepsilon_C^{k_A+k_B+\delta_A+\delta_B-k_C-\delta_C}$
$\leq \tilde{G}_C = (F_A(k_A - h_A + 1) + G_A)(F_B(k_B - h_B + 1) + G_B)$;

(c) $\varepsilon_C = \varepsilon_A \wedge \varepsilon_B = \tilde{\varepsilon}_C$.

(iv) *If $B(\varepsilon), \varepsilon \in (0, \varepsilon_0]$ is a pivotal $(h_B, k_B, \delta_B, G_B, \varepsilon_B)$-expansion, then there exists $\varepsilon_C \leq \varepsilon'_0 \leq \varepsilon_0$ such that $B(\varepsilon) \neq 0, \varepsilon \in (0, \varepsilon'_0]$, and $C(\varepsilon) = \frac{1}{B(\varepsilon)}, \varepsilon \in (0, \varepsilon'_0]$ is a pivotal $(h_C, k_C, \delta_C, G_C, \varepsilon_C)$-expansion with parameters h_C, k_C and coefficients $c_r, r = h_C, \ldots, k_C$ given in proposition* **(iv)** *of Lemma 2.2, and parameters δ_C, G_C, and ε_C given by formulas:*

(a) $\delta_C = \delta_B = \tilde{\delta}_C$;

(b) $G_C = (\frac{|b_{h_B}|}{2})^{-1}(G_B \sum_{h_C \leq j \leq k_C} |c_j|\varepsilon_C^{j+h_B}$
$+ \sum_{k_B - h_B < i+j, h_B \leq i \leq k_B, h_C \leq j \leq k_C} |b_i||c_j|\varepsilon_C^{i+j-k_B+h_B-\delta_B})$
$\leq \tilde{G}_C = (\frac{|b_{h_B}|}{2})^{-1}(G_B F_C(k_C - h_C + 1) + F_B F_C(k_B - h_B + 1)(k_C - h_C + 1))$
$= (\frac{|b_{h_B}|}{2})^{-1}(F_B(k_B - h_B + 1) + G_B)F_C(k_C - h_C + 1)$;

(c) $\varepsilon_C = \varepsilon_B \wedge \tilde{\varepsilon}'_B \geq \tilde{\varepsilon}_C = \varepsilon_B \wedge \tilde{\varepsilon}_B$, *where*

$$\tilde{\varepsilon}'_B = \Big(\frac{|b_{h_B}|}{2(\sum_{h_B < i \leq k_B} |b_i|\varepsilon_B^{i-h_B-\delta_B}) + G_B \varepsilon_B^{k_B-h_B})}\Big)^{\frac{1}{\delta_B}} \geq \tilde{\varepsilon}_B = \Big(\frac{|b_{h_B}|}{2(F_B(k_B-h_B)+G_B)}\Big)^{\frac{1}{\delta_B}}.$$

(v) *If $A(\varepsilon), \varepsilon \in (0, \varepsilon_0]$ is a $(h_A, k_A, \delta_A, G_A, \varepsilon_A)$-expansion, $B(\varepsilon), \varepsilon \in (0, \varepsilon_0]$ is a pivotal $(h_B, k_B, \delta_B, G_B, \varepsilon_B)$-expansion, then there exists $\varepsilon_D \leq \varepsilon'_0 \leq \varepsilon_0$ such that $B(\varepsilon) \neq 0, \varepsilon \in (0, \varepsilon'_0]$, and $D(\varepsilon) = \frac{A(\varepsilon)}{B(\varepsilon)}$ is a $(h_D, k_D, \delta_D, G_D, \varepsilon_D)$-expansion with parameters h_D, k_D and coefficients $d_r, r = h_D, \ldots, k_D$ given in proposition* **(v)** *of Lemma 2.2, and parameters $\delta_D, G_D, \varepsilon_D$ given by formulas:*

(a) $\delta_D = \delta_A \mathrm{I}(h_C + k_A < h_A + k_C) + (\delta_A \wedge \delta_C)\mathrm{I}(h_C + k_A = h_A + k_C)$
$+ \delta_C \mathrm{I}(h_A + k_C < h_C + k_A) \geq \tilde{\delta}_D = \delta_A \wedge \delta_C = \delta_A \wedge \delta_B$;

(b) $G_D = \sum_{k_D < i+j, h_A \leq i \leq k_A, h_C \leq j \leq k_C} |a_i||c_j|\varepsilon_D^{i+j-k_D-\delta_D}$
$+ G_A \sum_{h_C \leq j \leq k_C} |c_j|\varepsilon_D^{j+k_A+\delta_A-k_D-\delta_D} + G_C \sum_{h_A \leq i \leq k_A} |a_i|\varepsilon_D^{i+k_C+\delta_C-k_D-\delta_D}$
$+ G_A G_C \varepsilon_D^{k_A+k_C+\delta_A+\delta_C-k_D-\delta_D}$
$\leq \tilde{G}_D = (F_A(k_A - h_A + 1) + G_A)(F_C(k_C - h_C + 1) + G_C)$;

(c) $\varepsilon_D = \varepsilon_A \wedge \varepsilon_C = \tilde{\varepsilon}_D$,

where coefficients $c_r, r = h_C, \ldots, k_C$ and parameters $h_C, k_C, \delta_C, G_C, \varepsilon_C$ are given for the $(h_C, k_C, \delta_C, G_C, \varepsilon_C)$- expansion of function $C(\varepsilon) = \frac{1}{B(\varepsilon)}$ in the above proposition **(iv)**, *or by formulas:*

(d) $\delta_D = \delta_A \mathrm{I}(k_A - h_B < k_B - 2h_B + h_A) + (\delta_A \wedge \delta_B)\mathrm{I}(k_A - h_B = k_B - 2h_B + h_A) + \delta_B \mathrm{I}(k_A - h_B > k_B - 2h_B + h_A) \geq \tilde{\delta}_D = \delta_A \wedge \delta_B$;

(e) $G_D = (\frac{|b_{h_B}|}{2})^{-1}\big(\sum_{k_A \wedge (h_A + k_B - h_B) < i \leq k_A} |a_i| \varepsilon_D^{i - h_B - k_D - \delta_D} + G_A \varepsilon_D^{k_A + \delta_A - h_B - k_D - \delta_D} + \sum_{k_A \wedge (h_A + k_B - h_B) < i + j, h_B \leq i \leq k_B, h_D \leq j \leq k_D} |b_i||d_j| \varepsilon_D^{i + j - k_D - h_B - \delta_D} + G_B \sum_{h_D \leq j \leq k_D} |d_j| \varepsilon_D^{j + k_B + \delta_B - h_B - k_D - \delta_D}\big)$
$\leq \tilde{G}_D = (\frac{|b_{h_B}|}{2})^{-1}\big(F_A(k_A - k_A \wedge (h_A + k_B - h_B)) + G_A + F_B F_D (k_B - h_B + 1)(k_D - h_D + 1) + G_B F_D (k_D - h_D + 1)\big)$;

(f) $\varepsilon_D = \varepsilon_A \wedge \varepsilon_B \wedge \tilde{\varepsilon}'_B \geq \tilde{\varepsilon}_D = \varepsilon_A \wedge \varepsilon_B \wedge \tilde{\varepsilon}_B$, *where* $\tilde{\varepsilon}'_B$ *and* $\tilde{\varepsilon}_B$ *are given in relations* **(c)** *of proposition* **(iv)**.

Remark 2.8. Denominators in fractions representing parameters $\tilde{\varepsilon}'_B$ and $\tilde{\varepsilon}_B$ can take value 0. In this case, parameters $\tilde{\varepsilon}'_B$ and $\tilde{\varepsilon}_B$ take value ∞. In particular, this is, according to Remark 2.7, the case, if $B(\varepsilon) \equiv b \neq 0$.

Remark 2.9. Coefficients $\varepsilon_B, \varepsilon_C, \varepsilon_D \in (0, 1]$ are taken to nonnegative powers in all terms of the sums, which define parameters G_C, G_D, and $\tilde{\varepsilon}'_B$, in Lemma 2.7. This makes it possible to estimate the corresponding parameters G_C and G_D from above and parameter $\tilde{\varepsilon}'_B$ from below by the corresponding simpler expressions. In particular, the estimates by parameters $\tilde{G}_C$, $\tilde{G}_D$, and $\tilde{\varepsilon}_B$ are obtained by replacing coefficients $\varepsilon_B, \varepsilon_C$, and ε_D by 1.

Remark 2.10. Expansion $C(\varepsilon)$ given in propositions **(i)**–**(iv)** is also a $(h_C, k_C, \tilde{\delta}_C, \tilde{G}_C, \tilde{\varepsilon}_C)$-expansion with parameters $\tilde{\delta}_C, \tilde{G}_C$, and $\tilde{\varepsilon}_C$ given by simpler formulas than parameters δ_C, G_C, and ε_C. Also, expansion $D(\varepsilon)$ given in proposition **(v)** is a $(h_D, k_D, \tilde{\delta}_D, \tilde{G}_D, \tilde{\varepsilon}_D)$-expansion with parameters $\tilde{\delta}_D, \tilde{G}_D$, and $\tilde{\varepsilon}_D$ given by simpler formulas than parameters δ_D, G_D, and ε_D.

Let $A_m(\varepsilon) = a_{h_{A_m}, m}\varepsilon^{h_{A_m}} + \cdots + a_{k_{A_m}, m}\varepsilon^{k_{A_m}} + o_{A_m}(\varepsilon^{k_{A_m}}), \varepsilon \in (0, \varepsilon_0]$ be a $(h_{A_m}, k_{A_m}, \delta_{A_m}, G_{A_m}, \varepsilon_{A_m})$-expansion, for $m = 1, \ldots, N$.

Also, let us denote $F_{A_m} = \max_{h_{A_m} \leq i \leq k_{A_m}} |a_{i,m}|$, for $m = 1, \ldots, N$.

Lemma 2.7. *The above asymptotic expansions have the following multiple operational rules for computing remainders*:

(i) $B_n(\varepsilon) = A_1(\varepsilon) + \cdots + A_n(\varepsilon), \varepsilon \in (0, \varepsilon_0]$ *is, for every* $n = 1, \ldots, N$, *a* $(h_{B_n}, k_{B_n}, \delta_{B_n}, G_{B_n}, \varepsilon_{B_n})$*-expansion, with parameters* h_{B_n}, k_{B_n}, $n = 1, \ldots, N$ *and coefficients* $b_{h_{B_n}+l, n}, l = 0, \ldots, k_{B_n} - h_{B_n}, n = 1, \ldots, N$ *given in proposition* **(i)** *of Lemma 2.3, and parameters* $G_{B_n}, \delta_{B_n}, \varepsilon_{B_n}, n = 1, \ldots, N$ *given by formulas*:

(a) $\delta_{B_n} = \min_{m \in \mathbb{K}_n} \delta_{A_m} \geq \delta^*_n = \min_{1 \leq m \leq n} \delta_{A_m}$,
where $\mathbb{K}_n = \{m : 1 \leq m \leq n,\ k_{A_m} = \min(k_{A_1}, \ldots, k_{A_n})\}$;

(b) $G_{B_n} = \sum_{1 \leq i \leq n}\big(\sum_{k_{B_n} < j \leq k_{A_i}} |a_{j,i}| \varepsilon_{B_n}^{j - k_{B_n} - \delta_{B_n}} + G_{A_i} \varepsilon_{B_n}^{k_{A_i} + \delta_{A_i} - k_{B_n} - \delta_{B_n}}\big)$
$\leq \sum_{1 \leq i \leq n}\big(F_{A_i}(k_{A_i} - k_{B_n}) + G_{A_i}\big)$;

(c) $\varepsilon_{B_n} = \min(\varepsilon_{A_1}, \ldots, \varepsilon_{A_n})$.

(ii) *$C_n(\varepsilon) = A_1(\varepsilon) \times \cdots \times A_n(\varepsilon)$, $\varepsilon \in (0, \varepsilon_0]$ is, for $n = 1,\dots,N$, a $(h_{C_n}, k_{C_n}, \delta_{C_n}, G_{C_n}, \varepsilon_{C_N})$-expansion with parameters $h_{C_n}, k_{C_n}, n = 1,\dots,N$ and coefficients $c_{h_{C_n}+l,n}$, $l = 0,\dots,k_{C_n} - h_{C_n}, n = 1,\dots,N$ given in proposition* **(ii)** *of Lemma 2.3, and parameters $G_{C_n}, \delta_{C_n}, \varepsilon_{C_n}, n = 1,\dots,N$ given by formulas:*

(a) $\delta_{C_n} = \min_{m \in \mathbb{L}_n} \delta_{A_m} \geq \delta_n^*$, *where* $\mathbb{L}_n = \{m : 1 \leq m \leq n, (k_{A_m} + \sum_{1\leq r\leq n, r\neq m} h_{A_r}) = \min_{1\leq l\leq n}(k_{A_l} + \sum_{1\leq r\leq n, r\neq l} h_{A_r})\}$;

(b)
$$
\begin{aligned}
G_{C_n} = &\sum_{k_{C_n} < l_1+\cdots+l_n, h_{A_m} \leq l_m \leq k_{A_m}, 1\leq m\leq n} \prod_{1\leq m\leq n} |a_{l_m,m}| \varepsilon_{C_n}^{l_1+\cdots+l_n-k_{C_n}-\delta_{C_n}} \\
&+ \sum_{1\leq r\leq n} \sum_{1\leq m_1<\cdots<m_r\leq n} \Big(\prod_{m\neq m_1,\dots,m_r, 1\leq m\leq n} \big(\sum_{h_{A_m}\leq l\leq k_{A_m}} |a_{l,m}| \varepsilon_{C_n}^l \big) \\
&\times \prod_{1\leq i\leq r} G_{A_{m_i}} \varepsilon_{C_n}^{k_{A_{m_i}}+\delta_{A_{m_i}}-k_{C_n}-\delta_{C_n}} \Big) \\
&\leq \prod_{1\leq m\leq n} (F_{A_m}(k_{A_m} - h_{A_m} + 1) + G_{A_m}).
\end{aligned}
$$

(c) $\varepsilon_{C_n} = \min_{1\leq i\leq n} \varepsilon_{A_i}$.

(iii) *Parameters $\delta_{B_n}, G_{B_n}, \varepsilon_{B_n}, n = 1,\dots,N$ and $\delta_{C_n}, G_{C_n}, \varepsilon_{C_n}, n = 1,\dots,N$ in upper bounds for remainders in the asymptotic expansions, respectively, for functions $B_n(\varepsilon) = A_1(\varepsilon) + \cdots + A_n(\varepsilon)$, $n = 1,\dots,N$ and $C_n(\varepsilon) = A_1(\varepsilon) \times \cdots \times A_n(\varepsilon), n = 1,\dots,N$ are invariant with respect to any permutation, respectively, of summation and multiplication order in the above formulas.*

2.2.3 Algebraic Properties of Operational Rules for Laurent Asymptotic Expansions with Explicit Upper Bounds for Remainders

The summation and multiplication rules for computing of upper bounds for remainders given in propositions **(ii)** and **(iii)** of Lemma 2.6 possess the communicative property, but do not possess the associative and distributional properties.

Lemma 2.6 let us get an effective low bound for parameter δ_A for any $(h_A, k_A, \delta_A, G_A, \varepsilon_A)$-expansion $A(\varepsilon)$ obtained as the result of a finite sequence of operations (described in Lemma 2.6) performed over expansions from some finite set of such expansions.

Lemma 2.8. *The summation and multiplication operations for Laurent asymptotic expansions defined in Lemma 2.6 possess the following algebraic properties, which should be understood as equalities for the corresponding parameters of upper bounds for their remainders:*

(i) *The functional identity, $C(\varepsilon) \equiv A(\varepsilon) + B(\varepsilon) \equiv B(\varepsilon) + A(\varepsilon)$, implies that $\delta_C = \delta_{A+B} = \delta_{B+A}, G_C = G_{A+B} = G_{B+A}$ and $\varepsilon_C = \varepsilon_{A+B} = \varepsilon_{B+A}$.*

(ii) *The functional identity, $C(\varepsilon) \equiv A(\varepsilon) \cdot B(\varepsilon) \equiv B(\varepsilon) \cdot A(\varepsilon)$, implies that $\delta_C = \delta_{A\cdot B} = \delta_{B\cdot A}, G_C = G_{A\cdot B} = G_{B\cdot A}$ and $\varepsilon_C = \varepsilon_{A\cdot B} = \varepsilon_{B\cdot A}$.*

(iii) *If $A(\varepsilon)$ is $(h_A, k_A, \delta_A, G_A, \varepsilon_A)$-expansion obtained as the result of a finite sequence of operations (multiplication by a constant, summation, multiplication, and division) performed over $(h_{A_i}, k_{A_i}, \delta_{A_i}, G_{A_i}, \varepsilon_{A_i})$-expansions $A_i(\varepsilon), i = 1,\dots,N$, ac-*

cording to the rules presented in Lemmas 2.2 and 2.6, then $\delta_A \geq \delta_N^* = \min_{1\leq i\leq N}\delta_{A_i}$. *This makes it possible to rewrite* $A(\varepsilon)$ *as the* $(h_A, k_A, \delta_N^*, G_{A,N}^*, \varepsilon_A)$*-expansion, with parameter* $G_{A,N}^* = G_A\varepsilon_A^{\delta_A-\delta_N^*}$.

2.3 Proofs of Lemmas 2.1–2.8

In Section 2.3, we give proofs of Lemmas 2.1–2.8 omitting some known or obvious details. The chosen order of proofs has the sense because of the corresponding pairs of lemmas relate to the same functional identities, as Lemmas 2.1 and 2.5, or the same operational rules, as Lemmas 2.2 and 2.6 or 2.3 and 2.7, or analogous algebraic properties of the corresponding operational rules, as Lemmas 2.4 and 2.8. In each pair, the first lemma relates to Laurent asymptotic expansions with remainders given in the standard form, while the second lemma relates to Laurent asymptotic expansions with explicit upper bounds for remainders.

2.3.1 Lemmas 2.1 and 2.5

Let us, for the moment, use notations $A'(\varepsilon) = a'_{h'_A}\varepsilon^{h'_A} + \cdots + a'_{k'_A}\varepsilon^{k'_A} + o'_A(\varepsilon^{k'_A})$ and $A''(\varepsilon) = a''_{h''_A}\varepsilon^{h''_A} + \cdots + a''_{k''_A}\varepsilon^{k''_A} + o''_A(\varepsilon^{k''_A})$.

In the case $h'_A = h''_A$, proposition **(i)** of Lemma 2.1 is trivial. Let, for example, $h'_A < h''_A = h_A$. In this case, the assumption that proposition **(i)** does not hold, implies that there exists $h'_A \leq l < h_A$ such that $a'_k = 0$, for $h'_A \leq k < l$ and $a'_l \neq 0$. This implies that $e^{-l}A'(\varepsilon) \to a'_l \neq 0$ as $\varepsilon \to 0$, while $e^{-l}A''(\varepsilon) \to 0$ as $\varepsilon \to 0$. This contradicts to the initial identity $A(\varepsilon) \equiv A'(\varepsilon) \equiv A''(\varepsilon)$. Proposition **(i)** let us rewrite $A'(\varepsilon)$ and $A''(\varepsilon)$ in the more informative forms, $A'(\varepsilon) = a'_{h_A}\varepsilon^{h_A} + \cdots + a'_{k'_A}\varepsilon^{k'_A} + o'_A(\varepsilon^{k'_A})$ and $A''(\varepsilon) = a''_{h_A}\varepsilon^{h_A} + \cdots + a''_{k''_A}\varepsilon^{k''_A} + o''_A(\varepsilon^{k''_A})$. These representations imply that $\varepsilon^{-h_A}A'(\varepsilon) \to a'_{h_A}$ as $\varepsilon \to 0$ and $\varepsilon^{-h_A}A''(\varepsilon) \to a''_{h_A}$ as $\varepsilon \to 0$. These relations imply that $a_{h_A} = a'_{h'_A} = a''_{h''_A}$, since $A(\varepsilon) \equiv A'(\varepsilon) \equiv A''(\varepsilon)$. Also, $\varepsilon^{-h_A-1}(A'(\varepsilon) - a_{h_A}\varepsilon^{h_A}) \to a'_{h_A+1}$ as $\varepsilon \to 0$ and $\varepsilon^{-h_A-1}(A''(\varepsilon) - a_{h_A}\varepsilon^{h_A}) \to a''_{h_A+1}$ as $\varepsilon \to 0$. These relations imply that $a_{h_A+1} = a'_{h_A+1} = a''_{h_A+1}$, since $A(\varepsilon) - a_{h_A}\varepsilon^{h_A} \equiv A'(\varepsilon) - a_{h_A}\varepsilon^{h_A} \equiv A''(\varepsilon) - a_{h_A}\varepsilon^{h_A}$. By continuing the above procedure, we can prove that $a_l = a'_l = a''_l$, for $l = h_a \ldots, \tilde{k}_A = k'_A \wedge k''_A$, proposition **(ii)** of the lemma. By canceling the terms, $a_l\varepsilon^l \equiv a'_l\varepsilon^l \equiv a''_l\varepsilon^l, l = h_A, \ldots, \tilde{k}_A$ on the left- and right-hand sides of the initial identity $A(\varepsilon) = A'(\varepsilon) \equiv A''(\varepsilon)$, we get the identity, $\sum_{\tilde{k}_A < l \leq k'_A} a'_l\varepsilon^l + o'_A(\varepsilon^{k'_A}) \equiv \sum_{\tilde{k}_A < l \leq k''_A} a''_l\varepsilon^l + o''_A(\varepsilon^{k''_A})$, given in proposition **(v)**. This identity let us rewrite the Laurent asymptotic expansions $A(\varepsilon)$ in one of the two alternative forms $A(\varepsilon) \equiv A'(\varepsilon) \equiv \sum_{h_A \leq \tilde{k}_A} a_l\varepsilon^l + \sum_{\tilde{k}_A < l \leq k'_A} a'_l\varepsilon^l + o'_A(\varepsilon^{k'_A})$ or $A(\varepsilon) \equiv A''(\varepsilon) \equiv \sum_{h_A \leq \tilde{k}_A} a_l\varepsilon^l + \sum_{\tilde{k}_A < l \leq k''_A} a''_l\varepsilon^l + o''_A(\varepsilon^{k''_A})$. The first or second representation should be chosen as more informative if, respectively, $k'_A > k''_A$ or $k'_A < k''_A$. These representations coincide, if $k'_A = k''_A$. The above remark prove propositions **(iii)**–**(v)** of Lemma 2.1.

Lemma 2.5, in particular, relation (2.3) directly follows from formulas of remainders of the Laurent asymptotic expansion $A(\varepsilon)$ given in proposition **(v)** of Lemma 2.1 and priority rules for choice of parameters δ_A, G_A, and ε_A for explicit upper bounds for the above remainder, which are used in Lemma 2.5. □

2.3.2 Lemmas 2.2 and 2.6

Propositions **(i)** (the multiplication by a constant rule) of Lemmas 2.2 and 2.6 are obvious. Proposition **(ii)** (the summation rules) of Lemmas 2.2 and 2.6 can be obtained by simple accumulation of coefficients for different powers of ε and terms accumulated in the corresponding remainders, and, then, by using obvious upper bounds for absolute values of sums of terms accumulated in the corresponding remainders.

Proposition **(iii)** (the multiplication rule) of Lemma 2.2 can be proved by multiplication of the corresponding asymptotic expansions $A(\varepsilon)$ and $B(\varepsilon)$ and accumulation of coefficients for powers ε^l for $l = h_C, \ldots, k_C$ in their product,

$$
\begin{aligned}
C(\varepsilon) &= A(\varepsilon)B(\varepsilon) \\
&= (a_{h_A}\varepsilon^{h_A} + \cdots + a_{k_A}\varepsilon^{k_A} + o_A(\varepsilon^{k_A}))(b_{h_B}\varepsilon^{h_B} + \cdots + b_{k_B}\varepsilon^{k_B} + o_B(\varepsilon^{k_B})) \\
&= \sum_{h_C \le l \le k_C} \ \sum_{i+j=l, h_A \le i \le k_A, h_B \le j \le k_B} a_i b_j \varepsilon^l \\
&\quad + \sum_{k_C < i+j, h_A \le i \le k_A, h_B \le j \le k_B} a_i b_j \varepsilon^{i+j} + \sum_{h_B \le j \le k_B} b_j \varepsilon^j o_A(\varepsilon^{k_A}) \\
&\quad + \sum_{h_A \le i \le k_A} a_i \varepsilon^i o_B(\varepsilon^{k_B}) + o_A(\varepsilon^{k_A}) o_B(\varepsilon^{k_B}) = \sum_{h_C \le l \le k_C} c_l \varepsilon^l + o_C(\varepsilon^{k_C}),
\end{aligned} \tag{2.4}
$$

where

$$
\begin{aligned}
o_C(\varepsilon^{k_C}) = &\sum_{k_C < i+j, h_A \le i \le k_A, h_B \le j \le k_B} a_i b_j \varepsilon^{i+j} + \sum_{h_B \le j \le k_B} b_j \varepsilon^j o_A(\varepsilon^{k_A}) \\
&+ \sum_{h_A \le i \le k_A} a_i \varepsilon^i o_B(\varepsilon^{k_B}) + o_A(\varepsilon^{k_A}) o_B(\varepsilon^{k_B}).
\end{aligned} \tag{2.5}
$$

Obviously, $\frac{o_C(\varepsilon^{k_C})}{\varepsilon^{k_C}} \to 0$ as $\varepsilon \to 0$. It should be noted that the accumulation of coefficients for powers ε^l can be made in (2.4) only up to the maximal value $l = k_C = (k_A + h_B) \wedge (k_B + h_A)$, because of the presence in the expression for remainder $o_C(\varepsilon^{k_C})$ terms $b_{h_B}\varepsilon^{h_B} o_A(\varepsilon^{k_A})$ and $a_{h_A}\varepsilon^{h_A} o_B(\varepsilon^{k_B})$.

Also, relation (2.5) and Remark 2.9 readily imply relations **(a)**–**(c)**, which determines parameters $\delta_C, G_C, \varepsilon_C$ and $\tilde{\delta}_C, \tilde{G}_C, \tilde{\varepsilon}_C$ in proposition **(iii)** of Lemma 2.6, in particular, the following inequalities take place, for $\varepsilon \in (0, \varepsilon_C]$,

$$
\begin{aligned}
|\varepsilon^{-k_C-\delta_C} o_C(\varepsilon^{k_C})| \leq G_C = & \sum_{k_C < i+j, h_A \leq i \leq k_A, h_B \leq j \leq k_B} |a_i||b_j|\varepsilon_C^{i+j-k_C-\delta_C} \\
& + G_A \sum_{h_B \leq j \leq k_B} |b_j|\varepsilon_C^{j+k_A+\delta_A-k_C-\delta_C} + G_B \sum_{h_A \leq i \leq k_A} |a_i|\varepsilon_C^{i+k_B+\delta_B-k_C-\delta_C} \\
& + G_A G_B \varepsilon_C^{k_A+k_B+\delta_A+\delta_B-k_C-\delta_C} \\
& \leq \tilde{G}_C = F_A F_B (k_A - h_A + 1)(k_B - h_B + 1) \\
& \quad + G_A F_B (k_B - h_B + 1) + G_B F_A (k_A - h_A + 1) + G_A G_B \\
& \quad = (F_A (k_A - h_A + 1) + G_A)(F_B (k_B - h_B + 1) + G_B). \qquad (2.6)
\end{aligned}
$$

The assumptions of proposition **(iv)** in Lemma 2.2 imply that $\varepsilon^{-h_B} B(\varepsilon) \to b_{h_B} \neq 0$ as $\varepsilon \to 0$. This relation implies that there exists $0 < \varepsilon_0' \leq \varepsilon_0$ such that $B(\varepsilon) \neq 0$ for $\varepsilon \in (0, \varepsilon_0']$, and, thus, function $C(\varepsilon) = \frac{1}{B(\varepsilon)}$ is well defined for $\varepsilon \in (0, \varepsilon_0']$.

Note that $h_B \leq k_B$. The assumptions of proposition **(iv)** of Lemma 2.2 imply that, $\varepsilon^{h_B} C(\varepsilon) = (b_{h_B} + \cdots + b_{k_B} \varepsilon^{k_B - h_B} + o_B(\varepsilon^{h_B})\varepsilon^{-h_B})^{-1} \to b_{h_B}^{-1} = c_{h_C}$ as $\varepsilon \to 0$.

This relation means that function $\varepsilon^{h_B} C(\varepsilon)$ can be represented in the form $\varepsilon^{h_B} C(\varepsilon) = c_{h_C} + o(1)$, where $c_{h_C} = b_{h_B}^{-1}$, or, equivalently, that the following representation takes place, $C(\varepsilon) = c_{h_C} \varepsilon^{-h_B} + o_1(\varepsilon^{-h_B}), \varepsilon \in (0, \varepsilon_0']$, where $\frac{o_1(\varepsilon^{-h_B})}{\varepsilon^{-h_B}} \to 0$ as $\varepsilon \to 0$.

The latter two relations prove proposition **(iv)** of Lemma 2.2, for the case $h_B = k_B$. Indeed, these relations mean that function $C(\varepsilon) = \frac{1}{B(\varepsilon)}$ can be represented in the form of (h_C, k_C)-expansion with parameters $h_C = -h_B$, $k_C = k_B - 2h_B = -h_B = h_C$ and coefficient $c_{h_C} = b_{h_B}^{-1}$. Moreover, since $B(\varepsilon) \cdot C(\varepsilon) \equiv 1, 0 < \varepsilon \leq \varepsilon_0'$, remainder $c_1(\varepsilon)$ can be found from the following relation, $(b_{h_B} \varepsilon^{h_B} + o(\varepsilon^{h_B}))(c_{h_C} \varepsilon^{-h_B} + o_1(\varepsilon^{-h_B})) \equiv 1$ that yields formula, $o_1(\varepsilon^{-h_B}) = -\frac{c_{h_C} \varepsilon^{-h_B} o_B(\varepsilon^{h_B})}{b_{h_B} \varepsilon^{h_B} + o_B(\varepsilon^{h_B})}$.

This is formula **(c)** from proposition **(iv)** of Lemma 2.2, for the case $h_B = k_B$. Note that, in the case $h_B = k_B$, the above asymptotic expansion for function $C(\varepsilon)$ cannot be extended. Indeed, $\varepsilon^{h_B - 1} o_1(\varepsilon^{-h_B}) = \varepsilon^{h_B - 1}(C(\varepsilon) - c_{h_C} \varepsilon^{-h_B}) = -\frac{c_{h_C}}{b_{h_B} + o_B(\varepsilon^{h_B})\varepsilon^{-h_B}} \frac{o_B(\varepsilon^{h_B})\varepsilon^{-h_B}}{\varepsilon}$. The term $\frac{o_B(\varepsilon^{h_B})\varepsilon^{-h_B}}{\varepsilon}$ on the right-hand side in the latter relation has an uncertain asymptotic behavior as $\varepsilon \to 0$.

Let us now assume that $h_B + 1 \leq k_B$. In this case, the assumptions of proposition **(iv)** of Lemma 2.2 and the above asymptotic relations imply that, $\varepsilon^{h_B - 1} o_1(\varepsilon^{-h_B}) = \varepsilon^{h_B - 1}(C(\varepsilon) - c_{h_C} \varepsilon^{-h_B}) = (b_{h_B} + \cdots + b_{k_B} \varepsilon^{k_B - h_B} + o_B(\varepsilon^{k_B})\varepsilon^{-h_B})^{-1}(-b_{h_B+1} c_{h_C} - \cdots - b_{k_B} c_{h_C} \varepsilon^{k_B - h_B - 1} - o_B(\varepsilon^{k_B}) c_{h_C} \varepsilon^{-h_B - 1}) \to -b_{h_B}^{-1} b_{h_B+1} c_{h_C} = c_{h_C+1}$ as $\varepsilon \to 0$.

This relation means that function $\varepsilon^{h_B - 1} o_1(\varepsilon^{-h_B})$ can be represented in the form $\varepsilon^{h_B - 1} o_1(\varepsilon^{-h_B}) = c_{h_C+1} + o(1)$, where $c_{h_C+1} = -b_{h_B}^{-1} b_{h_B+1} c_{h_C}$, or, equivalently, that the following representation takes place, $C(\varepsilon) = c_{h_C} \varepsilon^{-h_B} + c_{h_C+1} \varepsilon^{-h_B+1} + o_2(\varepsilon^{-h_B+1}), \varepsilon \in (0, \varepsilon_0']$, where $\frac{o_2(\varepsilon^{-h_B+1})}{\varepsilon^{-h_B+1}} \to 0$ as $\varepsilon \to 0$.

The latter two relations prove proposition **(iv)** of Lemmas 2.2, for the case $h_B + 1 = k_B$. Indeed, these relations mean that function $C(\varepsilon)$ can be represented in the form of (h_C, k_C)-expansion with parameters $h_C = -h_B$, $k_C = k_B - 2h_B = -h_B + 1 = h_C + 1$ and coefficients $c_{h_C} = b_{h_B}^{-1}$, $c_{h_C+1} = -b_{h_B}^{-1} b_{h_B+1} c_{h_C}$. Moreover, since $B(\varepsilon) \cdot C(\varepsilon) \equiv 1$, the remainder $o_2(\varepsilon^{-h_B+1})$ can be found from the following relation: $(b_{h_B}\varepsilon^{h_B} + b_{h_B+1}\varepsilon^{h_B+1} + o_B(\varepsilon^{h_B+1}))(c_{h_C}\varepsilon^{-h_B} + c_{h_C+1}\varepsilon^{-h_B+1} + o_2(\varepsilon^{-h_B+1})) \equiv 1$. This yields formula, $o_2(\varepsilon^{-h_B+1}) = -\frac{b_{h_B+1}c_{h_C+1}\varepsilon^2 + (c_{h_C}\varepsilon^{-h_B} + c_{h_C+1}\varepsilon^{-h_B+1})o_B(\varepsilon^{h_B+1})}{b_{h_B}\varepsilon^{h_B} + b_{h_B+1}\varepsilon^{h_B+1} + o_B(\varepsilon^{h_B+1})}$.

This is formula **(c)** from proposition **(iv)** of Lemma 2.2, for the case $h_B + 1 = k_B$. Note that, in the case $h_B + 1 = k_B$, the above asymptotic expansion for function $C(\varepsilon)$ cannot be extended. Indeed, $\varepsilon^{h_B-2}o_2(\varepsilon) = \varepsilon^{h_B-2}(C(\varepsilon) - c_{h_C}\varepsilon^{-h_B} - c_{h_C+1}\varepsilon^{-h_B+1}) = -\frac{b_{h_B+1}c_{h_C+1} + c_{h_C+1}o_B(\varepsilon^{h_B+1})\varepsilon^{-h_B-1}}{b_{h_B} + b_{h_B+1}\varepsilon + o_B(\varepsilon^{h_B+1})\varepsilon^{-h_B}} - \frac{c_{h_C}}{b_{h_B} + b_{h_B+1}\varepsilon + o_B(\varepsilon^{h_B+1})\varepsilon^{-h_B}}\frac{o_B(\varepsilon^{h_B+1})\varepsilon^{-h_B-1}}{\varepsilon}$. The term $\frac{o_B(\varepsilon^{h_B+1})\varepsilon^{-h_B-1}}{\varepsilon}$ on the right-hand side in the latter relation has an uncertain asymptotic behavior as $\varepsilon \to 0$.

We can repeat the above arguments for the general case $h_B + n = k_B$, for any $n = 0, 1, \ldots$ and to prove that, in the case $h_B + n = k_B$, function $C(\varepsilon)$ can be represented in the form of (h_C, k_C)-expansion with parameters $h_C = -h_B, k_C = k_B - 2h_B = -h_B + n = h_C + n$ and coefficients $c_{h_C}, \ldots, c_{k_C}$ given in proposition **(iv)** of Lemma 2.2. Moreover, identity $B(\varepsilon) \cdot C(\varepsilon) \equiv 1, 0 < \varepsilon \leq \varepsilon_0'$, let us find the corresponding remainder $o_C(\varepsilon^{k_C})$ from the following relation:

$$(b_{h_B}\varepsilon^{h_B} + \cdots + b_{k_B}\varepsilon^{k_B} + o_B(\varepsilon^{k_B}))(c_{h_C}\varepsilon^{h_C} + \cdots + c_{h_C}\varepsilon^{k_C} + o_C(\varepsilon^{k_C})) \equiv 1. \quad (2.7)$$

Proposition **(iii)** of Lemma 2.2, applied to the product on the left-hand side in relation (2.7), permits to represent this product in the form of (h,k)-expansion with parameters $h = h_B + h_C = h_B - h_B = 0$ and $k = (k_B + h_C) \wedge (h_B + k_C) = (k_B - h_B) \wedge (k_B - 2h_B + h_B) = k_B - h_B$. By canceling coefficient for ε^l on the left- and right-hand sides in relation (2.7), for $l = 0, \ldots, k_B - h_B$, and then, by solving equation (2.7) with respect to the remainder $o_C(\varepsilon^{k_C})$, we get the formula for this remainder given in proposition **(iv)** of Lemma 2.2,

$$\begin{aligned} o_C(\varepsilon^{k_C}) &= -\frac{\sum_{k_B - h_B < i+j, h_B \leq i \leq k_B, h_C \leq j \leq k_C} b_i c_j \varepsilon^{i+j} + \sum_{h_C \leq j \leq k_C} c_j \varepsilon^j o_B(\varepsilon^{k_B})}{b_{h_B}\varepsilon^{h_B} + \cdots + b_{k_B}\varepsilon^{k_B} + o_B(\varepsilon^{k_B})} \\ &= -\frac{\sum_{k_B - h_B < i+j, h_B \leq i \leq k_B, h_C \leq j \leq k_C} b_i c_j \varepsilon^{i+j-h_B}}{b_{h_B} + \cdots + b_{k_B}\varepsilon^{k_B - h_B} + o_B(\varepsilon^{k_B})\varepsilon^{h_B}} \\ &\quad - \frac{\sum_{h_C \leq j \leq k_C} c_j \varepsilon^{j-h_B} o_B(\varepsilon^{k_B})}{b_{h_B} + \cdots + b_{k_B}\varepsilon^{k_B - h_B} + o_B(\varepsilon^{k_B})\varepsilon^{-h_B}}. \end{aligned} \quad (2.8)$$

The assumptions made in proposition **(iv)** of Lemma 2.6 imply that $B(\varepsilon) \neq 0$ and the following inequality holds for $0 < \varepsilon \leq \varepsilon_C$, where ε_C is given in proposition **(iv)** of Lemma 2.6:

$$\begin{aligned} &|b_{h_B+1}\varepsilon + \cdots + b_{k_B}\varepsilon^{k_B - h_B} + o_B(\varepsilon^{k_B})\varepsilon^{-h_B}| = \varepsilon^{\delta_B}(|b_{h_B+1}|\varepsilon^{1-\delta_B} + \cdots \\ &+ |b_{k_B}|\varepsilon^{k_B - h_B - \delta_B} + G_B\varepsilon^{k_B - h_B}) \leq \varepsilon^{\delta_B}(F_B(k_B - h_B) + G_B) \leq \frac{|b_{h_B}|}{2}, \end{aligned} \quad (2.9)$$

and, thus,

$$|b_{h_B} + b_{h_B+1}\varepsilon + \cdots + b_{k_B}\varepsilon^{k_B-h_B} + o_B(\varepsilon^{k_B})\varepsilon^{-h_B}| \geq |b_{h_B}|$$
$$-(|b_{h_B+1}|\varepsilon + \cdots + |b_{k_B}|\varepsilon^{k_B-h_B} + G_B\varepsilon^{k_B-h_B+\delta_B}) \geq \frac{|b_{h_B}|}{2} > 0. \quad (2.10)$$

The existence of ε_0' declared in proposition **(iv)** of Lemma 2.6 is obvious. For example, one can choose $\varepsilon_0' = \varepsilon_C$. It is also useful to note that formulas given in proposition **(iv)** of Lemma 2.6 imply that $\varepsilon_C = \varepsilon_B \wedge \tilde{\varepsilon}_B' \in (0, \varepsilon_0]$, since $\varepsilon_B \in (0, \varepsilon_0]$ and $\tilde{\varepsilon}_B' \in (0, \infty]$.

The assumptions made in proposition **(iv)** of Lemma 2.6, inequality (2.10), and Remark 2.8 imply that the following inequality holds, for $0 < \varepsilon \leq \varepsilon_C$:

$$\begin{aligned}
|\varepsilon^{-k_C-\delta_C} o_C(\varepsilon^{k_C})| &= |\varepsilon^{-k_B+2h_B-\delta_B} o_C(\varepsilon^{k_C})| \\
&\leq (\frac{|b_{h_B}|}{2})^{-1}(G_B \sum_{h_C \leq j \leq k_C} |c_j|\varepsilon_C^{j+h_B} \\
&\quad + \sum_{k_B-h_B<i+j, h_B\leq i\leq k_B, h_C\leq j\leq k_C} |b_i||c_j|\varepsilon_C^{i+j-k_B+h_B-\delta_B}) \\
&\leq \tilde{G}_C = (\frac{|b_{h_B}|}{2})^{-1}(G_B F_C(k_C - h_C + 1) \\
&\quad + F_B F_C(k_B - h_B + 1)(k_C - h_C + 1)) \\
&= (\frac{|b_{h_B}|}{2})^{-1}(G_B + F_B(k_B - h_B + 1))F_C(k_C - h_C + 1). \quad (2.11)
\end{aligned}$$

Inequality (2.11) proofs proposition **(iv)** of Lemma 2.6.

Propositions **(v)** of Lemmas 2.2 and 2.6 and relations **(a)**–**(c)** given in these propositions can be obtained by direct application, respectively, of propositions **(iii)** and **(iv)** of Lemmas 2.2 and 2.6 to the product $D(\varepsilon) = A(\varepsilon) \cdot \frac{1}{B(\varepsilon)}$.

Now, when it is already known that $D(\varepsilon) = A(\varepsilon) \cdot \frac{1}{B(\varepsilon)}$ is a (h_D, k_D)-expansion, with parameters $h_D = h_A - h_B$ and $k_D = (k_A - h_B) \wedge (k_B - 2h_B + h_A)$, multiplication of $D(\varepsilon)$ by $B(\varepsilon)$ yields the following relation holding for $\varepsilon \in (0, \varepsilon_0']$:

$$\begin{aligned}
A(\varepsilon) &= D(\varepsilon)B(\varepsilon) = a_{h_A}\varepsilon^{h_A} + \cdots + a_{h_A}\varepsilon^{k_A} + o_A(\varepsilon^{k_A}) \\
&= (d_{h_D}\varepsilon^{h_D} + \cdots + d_{h_D}\varepsilon^{k_D} + o_D(\varepsilon^{k_D}))(b_{h_B}\varepsilon^{h_B} + \cdots + b_{h_B}\varepsilon^{k_B} + o_B(\varepsilon^{k_B})). \quad (2.12)
\end{aligned}$$

By equating coefficients for powers ε^l for $l = h_D, \ldots, k_D$ on the left- and right-hand sides of the third equality in relation (2.12), we get alternative formulas **(e)** for coefficients $d_{h_d}, \ldots, d_{k_D}$ given in proposition **(v)** of Lemma 2.2.

Proposition **(iii)** of Lemma 2.2, applied to the product on the right-hand side in (2.12), permits to represent this product in the form of (h,k)-expansion with parameters $h = h_B + h_D = h_B + h_A - h_B = h_A$ and $k = (k_D + h_B) \wedge (k_B + h_D) = ((k_A - h_B) \wedge (k_B - 2h_B + h_A) + h_B) \wedge (k_B + h_A - h_B) = k_A \wedge (k_B + h_A - h_B)$. By canceling coefficient for ε^l on the left- and right-hand sides in relation (2.12), for

$l = h_A, \ldots, k_A \wedge (k_B + h_A - h_B)$, and then, by solving equation (2.12) with respect to the remainder $o_D(\varepsilon^{k_D})$, we get the formula **(f)** for this remainder given in proposition **(v)** of Lemma 2.2,

$$
\begin{aligned}
o_D(\varepsilon^{k_D}) &= \frac{\sum_{k_A \wedge (k_B+h_A-h_B) < l \le k_A} a_l \varepsilon^l + o_A(\varepsilon^{k_A})}{b_{h_B}\varepsilon^{h_B} + \cdots + b_{k_B}\varepsilon^{k_B} + o_B(\varepsilon^{k_B})} \\
&\quad - \frac{\sum_{k_A \wedge (k_B+h_A-h_B) < i+j, h_B \le i \le k_B, h_D \le j \le k_D} b_i d_j \varepsilon^{i+j}}{b_{h_B}\varepsilon^{h_B} + \cdots + b_{k_B}\varepsilon^{k_B} + o_B(\varepsilon^{k_B})} \\
&\quad - \frac{\sum_{h_D \le j \le k_D} d_j \varepsilon^j o_B(\varepsilon^{k_B})}{b_{h_B}\varepsilon^{h_B} + \cdots + b_{k_B}\varepsilon^{k_B} + o_B(\varepsilon^{k_B})} \\
&= \frac{\sum_{k_A \wedge (k_B+h_A-h_B) < l \le k_A} a_l \varepsilon^{l-h_B} + o_A(\varepsilon^{k_A})\varepsilon^{-h_B}}{b_{h_B} + \cdots + b_{k_B}\varepsilon^{k_B-h_B} + o_B(\varepsilon^{k_B})\varepsilon^{-h_B}} \\
&\quad - \frac{\sum_{k_A \wedge (k_B+h_A-h_B) < i+j, h_B \le i \le k_B, h_D \le j \le k_D} b_i d_j \varepsilon^{i+j-h_B}}{b_{h_B} + \cdots + b_{k_B}\varepsilon^{k_B-h_B} + o_B(\varepsilon^{k_B})\varepsilon^{-h_B}} \\
&\quad - \frac{\sum_{h_D \le j \le k_D} d_j \varepsilon^{j-h_B} o_B(\varepsilon^{k_B})}{b_{h_B} + \cdots + b_{k_B}\varepsilon^{k_B-h_B} + o_B(\varepsilon^{k_B})\varepsilon^{-h_B}}.
\end{aligned}
\tag{2.13}
$$

Inequality (2.10), the assumptions made in proposition **(v)** of Lemma 2.6 and Remark 2.8 finally imply that the following inequality holds, for $0 < \varepsilon \le \varepsilon_D$ given in relation **(f)** of this proposition:

$$
\begin{aligned}
|\varepsilon^{-k_D-\delta_D} o_D(\varepsilon^{k_D})| &\le G_D = \Big(\frac{|b_{h_B}|}{2}\Big)^{-1} \\
&\quad \times \Big(\sum_{k_A \wedge (h_A+k_B-h_B) < i \le k_A} |a_i| \varepsilon_D^{i-h_B-k_D-\delta_D} + G_A \varepsilon_D^{k_A+\delta_A-h_B-k_D-\delta_D} \\
&\quad + \sum_{k_A \wedge (h_A+k_B-h_B) < i+j, h_B \le i \le k_B, h_D \le j \le k_D} |b_i||d_j| \varepsilon_D^{i+j-k_D-h_B-\delta_D} \\
&\quad + G_B \sum_{h_D \le j \le k_D} |d_j| \varepsilon_D^{j+k_B+\delta_B-h_B-k_D-\delta_D} \Big) \\
&\le \tilde{G}_D = \Big(\frac{|b_{h_B}|}{2}\Big)^{-1} \big(F_A(k_A - k_A \wedge (h_A + k_B - h_B)) + G_A \\
&\quad F_B F_D (k_B - h_B + 1)(k_D - h_D + 1) + G_B F_D (k_D - h_D + 1)\big).
\end{aligned}
\tag{2.14}
$$

Inequality (2.14) completes the proof of proposition **(v)** of Lemma 2.6. □

Remark 2.11. Quantity $\frac{|b_{h_B}|}{2}$ can be replaced by quantity $\frac{|b_{h_B}|}{\alpha}$, for any $\alpha > 1$. Respectively, quantity $\frac{|b_{h_B}|}{2}$ can be replaced by quantity $\frac{|b_{h_B}|}{\beta}$, where $\beta = \frac{\alpha}{\alpha-1}$, on the right-hand side of inequality (2.10) and, in sequel, in inequalities (2.11) and (2.14). This makes it possible to modify propositions **(iv)** and **(v)** of Lemma 1.6 by replacing factor $\frac{|b_{h_B}|}{2}$ by new factor $\frac{|b_{h_B}|}{\alpha}$ in the expression for parameters $\tilde{\varepsilon}'_B, \tilde{\varepsilon}_B$ and by factor $\frac{|b_{h_B}|}{\beta}$ in the expression for parameters $G_C, \tilde{G}_C, G_D, \tilde{G}_D$. The choice of a smaller value for parameter α diminishes values of parameters $\tilde{\varepsilon}'_B, \tilde{\varepsilon}_B$ and $G_C, \tilde{G}_C, G_D, \tilde{G}_D$.

2.3.3 Lemmas 2.3 and 2.7

Lemma 2.3 is a direct corollary of Lemma 2.2.

Proofs of propositions **(i)** and **(ii)** in Lemma 2.7 are analogous to proofs of propositions **(i)** and **(ii)** in Lemma 2.6. Proposition **(iii)** of Lemma 2.7 is obvious. □

2.3.4 Lemmas 2.4 and 2.8

The first two identities for Laurent asymptotic expansions given in proposition **(i)** of Lemma 2.4 are obvious. The third identity given in this proposition follows in an obvious way from proposition **(i)** of Lemma 2.2. By applying propositions **(iii)** and **(iv)** of Lemma 2.2 to the product $C(\varepsilon) = A(\varepsilon) \cdot A(\varepsilon)^{-1}$, we get parameters $h_C = h_{A \cdot A^{-1}} = h_A - h_A = 0, k_C = k_{A \cdot A^{-1}} = (k_A - h_A) \wedge (k_A - 2h_A + h_A) = k_A - h_A$ and coefficients $c_n = \mathrm{I}(n = 0), n = 0, \ldots, k_C$. Also, relations (2.7) and (2.8) imply that the elimination identity $A(\varepsilon) \cdot A(\varepsilon)^{-1} \equiv 1$ holds, since the remainder of Laurent asymptotic expansion for function $A(\varepsilon)^{-1}$ is given by formula **(c)** from proposition **(iv)** of Lemma 2.2. Propositions **(ii)** and **(iii)** of Lemma 2.4 in the parts concerned commutative property of summation and multiplication operations follow from, respectively, propositions **(ii)** and **(iii)** of Lemma 2.2.

Let $D(\varepsilon) = (A(\varepsilon) + B(\varepsilon)) + C(\varepsilon) = A(\varepsilon) + (B(\varepsilon) + C(\varepsilon))$. Using propositions **(ii)** of Lemma 2, we get, $h_D = h_{(A+B)+C} = (h_A \wedge h_B) \wedge h_C = h_A \wedge (h_B \wedge h_C) = h_{A+(B+C)}$ and $k_D = k_{(A+B)+C} = (k_A \wedge k_B) \wedge k_C = k_A \wedge (k_B \wedge k_C) = k_{A+(B+C)}$. These relations and Lemma 2.1 imply equalities for the corresponding coefficients and remainders, for the asymptotic expansions of functions $(A(\varepsilon) + B(\varepsilon)) + C(\varepsilon)$ and $A(\varepsilon) + (B(\varepsilon) + C(\varepsilon))$. The above remarks prove proposition **(ii)** of Lemma 2.4 in the part concerned with the associative property of summation operation for Laurent asymptotic expansions.

Let $D(\varepsilon) = (A(\varepsilon) \cdot B(\varepsilon)) \cdot C(\varepsilon) = A(\varepsilon) \cdot (B(\varepsilon) \cdot C(\varepsilon))$. Using propositions **(iii)** of Lemma 2.2, we get, $h_D = h_{(A \cdot B) \cdot C} = h_{A \cdot B} + h_C = h_A + h_B + h_C = h_A + h_{B \cdot C} = h_{A \cdot (B \cdot C)}$ and $k_D = k_{(A \cdot B) \cdot C} = (k_{A \cdot B} + h_C) \wedge (k_C + h_{A \cdot B}) = ((k_A + h_B) \wedge (k_B + h_A)) + h_C) \wedge (k_C + (h_A + h_B)) = (k_A + h_B + h_C) \wedge (k_B + h_A + h_C) \wedge (k_C + h_A + h_B) = (k_A + (h_B + h_C)) \wedge ((k_B + h_C) \wedge (k_C + h_B)) + h_A) = (k_A + h_{B \cdot C}) \wedge (k_{B \cdot C} + h_A) = k_{A \cdot (B \cdot C)}$. These relations and Lemma 2.1 imply equalities for the corresponding coefficients and remainders, for the asymptotic expansions of functions $(A(\varepsilon) \cdot B(\varepsilon)) \cdot C(\varepsilon)$ and $A(\varepsilon) \cdot (B(\varepsilon) \cdot C(\varepsilon))$. The above remarks prove proposition **(iii)** of Lemma 2.4 in the part concerned with the associative property of multiplication operation for Laurent asymptotic expansions.

Let $D(\varepsilon) = (A(\varepsilon) + B(\varepsilon)) \cdot C(\varepsilon) = A(\varepsilon) \cdot C(\varepsilon) + B(\varepsilon) \cdot C(\varepsilon)$. Using propositions **(ii)** and **(iii)** of Lemma 2, we get, $h_D = h_{(A+B) \cdot C} = h_{A+B} + h_C = h_A \wedge h_B + h_C = (h_A + h_C) \wedge (h_B + h_C) = h_{A \cdot C} \wedge h_{B \cdot C} = h_{A \cdot C + B \cdot C}$ and $k_D = k_{(A+B) \cdot C} = (k_{A+B} + h_C) \wedge (k_C + h_{A+B}) = (k_A \wedge k_B + h_C) \wedge (k_C + h_A \wedge h_B) = (k_A + h_C) \wedge (k_B + h_C) \wedge (k_C + h_A) \wedge (k_C + h_B) = ((k_A + h_C) \wedge (k_C + h_A)) \wedge ((k_B + h_C) \wedge (k_C + h_B)) = k_{A \cdot C} \wedge k_{B \cdot C} = k_{A \cdot C + B \cdot C}$.

These relations and Lemma 2.1 imply equalities for the corresponding coefficients and remainders, for the asymptotic expansions of functions $(A(\varepsilon)+B(\varepsilon))\cdot C(\varepsilon)$ and $A(\varepsilon)\cdot C(\varepsilon)+B(\varepsilon)\cdot C(\varepsilon)$. The above remarks prove proposition **(iv)** of Lemma 2.4 concerned with the distributive property of summation and multiplication operations for Laurent asymptotic expansions.

Propositions **(i)** and **(ii)** of Lemma 2.8 readily follow from, respectively, propositions **(ii)** and **(iii)** of Lemma 2.6. Finally, proposition **(iii)** of Lemma 2.8 follows from relations $\delta_{cA} = \delta_A$, $\delta_{A+B}, \delta_{A\cdot B}$, $\delta_{A/B} \geq \delta_A \wedge \delta_B$ and $\delta_{A_1+\cdots+A_N}$, $\delta_{A_1\times\cdots\times A_N} \geq \min_{1\leq m\leq N} \delta_{A_m}$, given, respectively, in Lemmas 2.6 and 2.7. □

Chapter 3
Asymptotic Expansions for Moments of Hitting Times for Nonlinearly Perturbed Semi-Markov Processes

In Chapter 3, we introduce a model of perturbed semi-Markov processes, formulate basic perturbation conditions, describe a one-step time-space screening procedure of phase space reduction for perturbed semi-Markov processes, introduce hitting times, and prove an invariant property of them with respect to the procedure of phase space reduction. We, also, present algorithms for re-calculation of asymptotic expansions for transition characteristics of nonlinearly perturbed semi-Markov processes with reduced phase spaces and algorithms for sequential reduction of phase space for semi-Markov processes and construction of Laurent asymptotic expansions, without and with explicit upper bounds for remainders, for power moment of hitting times.

3.1 Perturbed Semi-Markov Processes

In Section 3.1 we introduce semi-Markov processes and present some basic facts concerned these processes. The corresponding proofs can be found, for example, in the books by Silvestrov (1980), Shurenkov (1989), Limnios and Oprişan (2001), Janssen and Manca (2006), Howard (2007), Gyllenberg and Silvestrov (2008), Harlamov (2008), and Grabski (2015). We, also, formulate the main perturbation conditions for nonlinearly perturbed semi-Markov processes.

3.1.1 Perturbed Semi-Markov Processes

Let $\mathbb{X} = \{1,\dots,N\}$ and $(\eta_{\varepsilon,n}, \kappa_{\varepsilon,n}), n = 0, 1, \dots$ be, for every $\varepsilon \in (0, \varepsilon_0]$, a Markov renewal process, i.e., a homogeneous Markov chain with the phase space $\mathbb{X} \times [0, \infty)$, an initial distribution $\bar{p}_\varepsilon = \langle p_{\varepsilon,i} = \mathsf{P}\{\eta_{\varepsilon,0} = i, \kappa_{\varepsilon,0} = 0\} = \mathsf{P}\{\eta_{\varepsilon,0} = i\}, i \in \mathbb{X}\rangle$, and transition probabilities defined for $(i,s),(j,t) \in \mathbb{X} \times [0,\infty), n \geq 1$,

D. Silvestrov, S. Silvestrov, *Nonlinearly Perturbed Semi-Markov Processes*,
SpringerBriefs in Probability and Mathematical Statistics,
DOI 10.1007/978-3-319-60988-1_3

$$\begin{aligned} Q_{\varepsilon,ij}(t) &= \mathsf{P}\{\eta_{\varepsilon,n} = j, \kappa_{\varepsilon,n} \leq t / \eta_{\varepsilon,n-1} = i, \kappa_{\varepsilon,n-1} = s\} \\ &= \mathsf{P}\{\eta_{\varepsilon,1} = j, \kappa_{\varepsilon,1} \leq t / \eta_{\varepsilon,0} = i\}. \end{aligned} \tag{3.1}$$

In this case (where the transition probabilities do not depend on s), the random sequence $\eta_{\varepsilon,n}, n = 0, 1, \ldots$ is also a homogeneous (embedded) Markov chain with the phase space $\mathbb{X}$ and the transition probabilities, defined for $i, j \in \mathbb{X}$,

$$p_{ij}(\varepsilon) = \mathsf{P}\{\eta_{\varepsilon,1} = j / \eta_{\varepsilon,0} = i\} = Q_{\varepsilon,ij}(\infty). \tag{3.2}$$

The following communication condition plays an important role in what follows:

A: There exist sets $\mathbb{Y}_i \subseteq \mathbb{X}, i \in \mathbb{X}$ such that: **(a)** probabilities $p_{ij}(\varepsilon) > 0$, $j \in \mathbb{Y}_i$, $i \in \mathbb{X}$, for $\varepsilon \in (0, \varepsilon_0]$; **(b)** probabilities $p_{ij}(\varepsilon) = 0$, $j \in \overline{\mathbb{Y}}_i$, $i \in \mathbb{X}$, for $\varepsilon \in (0, \varepsilon_0]$; **(c)** there exists, for every pair of states $i, j \in \mathbb{X}$, an integer $n_{ij} \geq 1$ and a chain of states $i = l_{ij,0}, l_{ij,1}, \ldots, l_{ij,n_{ij}} = j$ such that $l_{ij,1} \in \mathbb{Y}_{l_{ij,0}}, \ldots, l_{ij,n_{ij}} \in \mathbb{Y}_{l_{ij,n_{ij}-1}}$.

We refer to sets $\mathbb{Y}_i, i \in \mathbb{X}$ as transition sets. Condition **A** implies that all sets $\mathbb{Y}_i \neq \emptyset, i \in \mathbb{X}$.

Conditions **A (a)** and **(b)** mean that the communicative structure of the phase space $\mathbb{X}$ is the same for every $\varepsilon \in (0, \varepsilon_0]$.

Moreover, conditions **A (a)** and **(c)** imply that the phase space $\mathbb{X}$ of Markov chain $\eta_{\varepsilon,n}$ is one class of communicative states, for every $\varepsilon \in (0, \varepsilon_0]$.

Indeed, these conditions imply that, for every pair of states $i, j \in \mathbb{X}$, there exists a chain of states $i = l_{ij,0}, l_{ij,1}, \ldots, l_{ij,n_{ij}} = j$ with the corresponding chain probability, $\prod_{n=1}^{n_{ij}} p_{l_{ij,n-1} l_{ij,n}}(\varepsilon) > 0$, for every $\varepsilon \in (0, \varepsilon_0]$.

We also assume that the following regularity condition excluding instant transitions holds:

B: $Q_{\varepsilon,ij}(0) = 0, \ i, j \in \mathbb{X}$, for every $\varepsilon \in (0, \varepsilon_0]$.

Let us now introduce a semi-Markov process,

$$\eta_\varepsilon(t) = \eta_{\varepsilon,\nu_\varepsilon(t)}, \ t \geq 0, \tag{3.3}$$

where $\nu_\varepsilon(t) = \max(n \geq 0 : \zeta_{\varepsilon,n} \leq t)$ is the number of jumps in the time interval $[0, t]$, for $t \geq 0$, and $\zeta_{\varepsilon,n} = \kappa_{\varepsilon,1} + \cdots + \kappa_{\varepsilon,n}$, $n = 0, 1, \ldots$, are the sequential moments of jumps, for the semi-Markov process $\eta_\varepsilon(t)$.

The random variables $\kappa_{\varepsilon,1}, \kappa_{\varepsilon,2}, \ldots$ and $\eta_{\varepsilon,1}, \eta_{\varepsilon,2}, \ldots$ are, respectively, transition times (alternatively called by sojourn times or inter-jump times) and states at moments of jumps, for the semi-Markov process $\eta_\varepsilon(t)$.

Note that the regularity condition **B** implies that the random variables $\zeta_{\varepsilon,n} \xrightarrow{\mathsf{P}} \infty$ as $n \to \infty$ and, thus, $\nu_\varepsilon(t) < \infty$ with probability 1, for every $t \geq 0$.

If $Q_{\varepsilon,ij}(t) = \mathrm{I}(t \geq 1) p_{ij}(\varepsilon), t \geq 0, i, j \in \mathbb{X}$, then $\eta_\varepsilon(t) = \eta_{\varepsilon,[t]}, t \geq 0$ is a discrete time homogeneous Markov chain embedded in continuous time.

If $Q_{\varepsilon,ij}(t) = (1 - e^{-\lambda_i(\varepsilon)t}) p_{ij}(\varepsilon), t \geq 0, i, j \in \mathbb{X}$ (here, $0 < \lambda_i(\varepsilon) < \infty, i \in \mathbb{X}$), then $\eta_\varepsilon(t), t \geq 0$ is a continuous time homogeneous Markov chain.

3.1.2 Hitting Times for Semi-Markov Processes

Let us introduce power moments of transition times,

$$e_{ij}(k,\varepsilon) = \mathsf{E}_i \kappa_{\varepsilon,1}^k \mathrm{I}(\eta_{\varepsilon,1} = j) = \int_0^\infty t^k Q_{\varepsilon,ij}(dt),\ k = 0,1,2,\ldots,\ i,j \in \mathbb{X}. \tag{3.4}$$

Here and henceforth, notations P_i and E_i are used for conditional probabilities and expectations under the condition, $\eta_\varepsilon(0) = i$.

It is useful to note that the transition probabilities $p_{ij}(\varepsilon) = e_{ij}(0,\varepsilon), i,j \in \mathbb{X}$ can be interpreted as moments of transition times of order 0.

We assume that the following condition holds for some integer $d \geq 1$:

$\mathbf{C}_d$: $e_{ij}(d,\varepsilon) < \infty,\ i,j \in \mathbb{X}$, for $\varepsilon \in (0,\varepsilon_0]$.

Conditions **A (a)–(b)**, **B**, and $\mathbf{C}_d$ imply that, for every $\varepsilon \in (0,\varepsilon_0]$, expectations $e_{ij}(k,\varepsilon) > 0$, for $k = 1,\ldots,d, j \in \mathbb{Y}_i, i \in \mathbb{X}$, and $e_{ij}(k,\varepsilon) = 0$, for $k = 1,\ldots,d, j \in \overline{\mathbb{Y}}_i, i \in \mathbb{X}$.

In the case of discrete time Markov chain, $e_{ij}(k,\varepsilon) = p_{ij}(\varepsilon),\ k \geq 1,\ i,j \in \mathbb{X}$.

In the case of continuous time Markov chain, $e_{ij}(k,\varepsilon) = \frac{k!}{\lambda_i^k(\varepsilon)} p_{ij}(\varepsilon),\ k \geq 1$ $i,j \in \mathbb{X}$.

In both cases, condition $\mathbf{C}_d$ holds for any $d \geq 1$.

Let us define hitting times, which are random variables given by the following relation, for $j \in \mathbb{X}$:

$$\tau_{\varepsilon,j} = \sum_{n=1}^{\nu_{\varepsilon,j}} \kappa_{\varepsilon,n},\ \text{where}\ \nu_{\varepsilon,j} = \min(n \geq 1 : \eta_{\varepsilon,n} = j). \tag{3.5}$$

Let us denote,

$$E_{ij}(k,\varepsilon) = \mathsf{E}_i \tau_{\varepsilon,j}^k,\ k = 0,\ldots,d, i,j \in \mathbb{X}. \tag{3.6}$$

As is known (see, for example, Silvestrov (1980) and Gyllenberg and Silvestrov (2008)), conditions **A**, **B**, and $\mathbf{C}_d$ imply that, for every $i,j \in \mathbb{X}$ and $\varepsilon \in (0,\varepsilon_0]$,

$$0 < E_{ij}(d,\varepsilon) < \infty. \tag{3.7}$$

Note that, $E_{ij}(0,\varepsilon) = 1, i,j \in \mathbb{X}$.

Taking into account that $\zeta_{\varepsilon,1}$ is Markov time for the Markov renewal process $(\eta_{\varepsilon,n}, \kappa_{\varepsilon,n})$, we can write down the following stochastic equality, for every $i,j \in \mathbb{X}$:

$$\tau_{\varepsilon,ij} \stackrel{d}{=} \kappa_{\varepsilon,i,1} I(\eta_{\varepsilon,i,1} = j) + \sum_{r \neq j} (\kappa_{\varepsilon,i,1} + \tau_{\varepsilon,rj}) I(\eta_{\varepsilon,i,1} = r), \tag{3.8}$$

where (a) $\tau_{\varepsilon,ij}$ is a nonnegative random variable, which has distribution $\mathsf{P}\{\tau_{\varepsilon,ij} \leq t\} = \mathsf{P}_i\{\tau_{\varepsilon,j} \leq t\}, t \geq 0$, for every $i,j \in \mathbb{X}$; (b) $(\eta_{\varepsilon,i,1}, \kappa_{\varepsilon,i,1})$ is a random vector, which takes values in space $\mathbb{X} \times [0,\infty)$ and has distribution $\mathsf{P}\{\eta_{\varepsilon,i,1} = j, \kappa_{\varepsilon,i,1} \leq t\} =$

$\mathsf{P}_i\{\eta_{\varepsilon,1} = j, \kappa_{\varepsilon,1} \le t\} = Q_{\varepsilon,ij}(t), j \in \mathbb{X}, t \ge 0$, for every $i \in \mathbb{X}$; (c) the random variable $\tau_{\varepsilon,rj}$ and the random vector $(\eta_{\varepsilon,i,1}, \kappa_{\varepsilon,i,1})$ are independent, for every $r, j, i \in \mathbb{X}$.

Here and henceforth, symbol $\overset{d}{=}$ means that random variables on the right- and left-hand sides of the corresponding stochastic equality have the same distribution.

Let us denote $G_{\varepsilon,ij}(t) = \mathsf{P}_i\{\tau_{\varepsilon,j} \le t\}$. Then relation (3.8) is (under assumptions (a)–(c) made for the random variables appearing in this relation) equivalent to the following relation (which follows from the Markov property of the underlying semi-Markov process at the moment of first jump):

$$G_{\varepsilon,ij}(t) = Q_{\varepsilon,ij}(t) + \sum_{r \neq j} \int_0^t G_{\varepsilon,rj}(t-s) Q_{\varepsilon,ir}(ds),\ t \ge 0, i, j \in \mathbb{X}. \tag{3.9}$$

By taking moments of the order k in stochastic relation (3.8) we get, for every $j \in \mathbb{X}$, the following system of linear equations for the moments $E_{ij}(k,\varepsilon), i \in \mathbb{X}$, for $k = 1, \ldots, d$:

$$\begin{cases} E_{ij}(k,\varepsilon) = e_{ij}(k,\varepsilon) + \sum_{r \neq j} \sum_{l=0}^{k-1} \binom{k}{l} e_{ir}(k-l,\varepsilon) E_{rj}(l,\varepsilon) \\ \qquad + \sum_{r \neq j} p_{ir}(\varepsilon) E_{rj}(k,\varepsilon),\ i \in \mathbb{X}. \end{cases} \tag{3.10}$$

Systems of linear equations given in relation (3.10) should be solved recurrently, for $k = 1, 2 \ldots, d$. Indeed, the expressions for free terms in system for moments $E_{ij}(k,\varepsilon), i \in \mathbb{X}$ include moments $E_{rj}(l,\varepsilon), r \in \mathbb{X}, l = 1, \ldots, k-1$.

Systems of linear equations (3.10) have the same matrix $\|p_{ir}(\varepsilon)\mathrm{I}(r \neq j)\|$ of coefficients, for $k = 1, \ldots, d$. This matrix has non-zero determinant.

Thus, moments $E_{ij}(k,\varepsilon), i \in \mathbb{X}$ are, for every $k = 1, \ldots, d$, the unique solution of the above system of linear equations (see, for example, Gyllenberg and Silvestrov (2008)).

Let us denote,

$$e_i(\varepsilon) = \mathsf{E}_i \kappa_{\varepsilon,1} = \sum_{j \in \mathbb{X}} e_{ij}(1,\varepsilon),\ i \in \mathbb{X}. \tag{3.11}$$

Conditions **B** and $\mathbf{C}_1$ imply that, for every $i \in \mathbb{X}$ and $\varepsilon \in (0, \varepsilon_0]$,

$$0 < e_i(\varepsilon) < \infty. \tag{3.12}$$

The free terms for the system of linear equations for expectations $E_{ij}(1,\varepsilon), i \in \mathbb{X}$ have the form, $e_{ij}(1,\varepsilon) + \sum_{r \neq j} e_{ir}(1,\varepsilon) E_{rj}(0,\varepsilon) = e_{ij}(1,\varepsilon) + \sum_{r \neq j} e_{ir}(1,\varepsilon) = e_i(\varepsilon)$, for $i \in \mathbb{X}$, and, thus, system (3.10) takes, in this case, the following form:

$$\{E_{ij}(1,\varepsilon) = e_i(\varepsilon) + \sum_{r \neq j} p_{ir}(\varepsilon) E_{rj}(1,\varepsilon),\ i \in \mathbb{X}. \tag{3.13}$$

3.1.3 Stationary Distributions for Semi-Markov Processes

Let us first recall some well-known facts concerned ergodic properties of Markov chains. The relations presented below can be found, for example, in Feller (1968), Limnios and Oprişan (2001), and Gyllenberg and Silvestrov (2008).

Condition **A** implies that, for every $\varepsilon \in (0, \varepsilon_0]$, the Markov chain $\eta_{\varepsilon,n}$ is ergodic. Its stationary distribution $\bar{\rho}(\varepsilon) = \langle \rho_1(\varepsilon), \dots, \rho_N(\varepsilon) \rangle$ is given by the ergodic relation,

$$\bar{\mu}_{\varepsilon,n,i} = \frac{1}{n}\sum_{k=1}^{n} \mathrm{I}(\eta_{\varepsilon,k-1} = i) \stackrel{a.s.}{\longrightarrow} \rho_i(\varepsilon) > 0 \text{ as } n \to \infty, \text{ for } i \in \mathbb{X}. \tag{3.14}$$

This ergodic relation holds for any initial distribution $\bar{p}^{(\varepsilon)}$ and the stationary distribution $\bar{\rho}(\varepsilon) = \langle \rho_1(\varepsilon), \dots, \rho_N(\varepsilon) \rangle$ does not depend on the initial distribution.

Also, two useful formulas take place,

$$\rho_i(\varepsilon) = \frac{1}{\mathsf{E}_i \nu_{\varepsilon,i}}, i \in \mathbb{X}, \tag{3.15}$$

The stationary probabilities $\rho_i(\varepsilon), i \in \mathbb{X}$ are the unique solution for the system of linear equations,

$$\left\{ \rho_i(\varepsilon) = \textstyle\sum_{j\in\mathbb{X}} \rho_j(\varepsilon) p_{ji}(\varepsilon), j \in \mathbb{X}, \ \sum_{i\in\mathbb{X}} \rho_i(\varepsilon) = 1. \right. \tag{3.16}$$

Analogous to (3.14)–(3.15) relations also take place for semi-Markov processes.

Conditions **A**, **B**, and $\mathbf{C}_1$ imply that, for every $\varepsilon \in (0, \varepsilon_0]$, the semi-Markov process $\eta_\varepsilon(t)$ is ergodic. Its stationary distribution $\bar{\pi}(\varepsilon) = \langle \pi_1(\varepsilon), \dots, \pi_N(\varepsilon) \rangle$ is given by the following ergodic relation (see, for example, Silvestrov (1980), Shurenkov (1989) and Limnios and Oprişan (2001)),

$$\bar{\mu}_{\varepsilon,i}(t) = \frac{1}{t}\int_0^t I(\eta_\varepsilon(s) = i)ds \stackrel{a.s.}{\longrightarrow} \pi_i(\varepsilon) > 0 \text{ as } t \to \infty, \text{ for } i \in \mathbb{X}. \tag{3.17}$$

The ergodic relation (3.17) holds for any initial distribution $\bar{p}^{(\varepsilon)}$ and the stationary distribution $\bar{\pi}(\varepsilon)$ does not depend on the initial distribution.

The following relation, which connect expected return times and stationary probabilities for semi-Markov processes, plays an important role in what follows (see, for example, Silvestrov (1980), Shurenkov (1989), Limnios and Oprişan (2001), and Janssen and Manca (2006)),

$$\pi_i(\varepsilon) = \frac{e_i(\varepsilon)}{E_{ii}(1,\varepsilon)}, \ i \in \mathbb{X}. \tag{3.18}$$

The stationary probabilities $\pi_i(\varepsilon), i \in \mathbb{X}$ are the unique solution for the system of linear equations,

$$\left\{ \pi_i(\varepsilon) e_i^{-1}(\varepsilon) = \textstyle\sum_{j\in\mathbb{X}} \pi_j(\varepsilon) e_j^{-1}(\varepsilon) p_{ji}(\varepsilon), j \in \mathbb{X}, \ \sum_{i\in\mathbb{X}} \pi_i(\varepsilon) = 1. \right. \tag{3.19}$$

Indeed, the following formula (see, for example, Silvestrov (1980) or Gyllenberg and Silvestrov (2008)) takes place, $E_{ii}(1,\varepsilon) = \sum_{j\in\mathbb{X}} e_j(\varepsilon)\mathsf{E}_i \nu_{\varepsilon,ij}$, where $\nu_{\varepsilon,ij} = \sum_{k=1}^{\nu_{\varepsilon,i}} \mathrm{I}(\eta_{\varepsilon,k-1} = j), i,j \in \mathbb{X}$. As well known, $\mathsf{E}_i \nu_{\varepsilon,ij} = \frac{\rho_j(\varepsilon)}{\rho_i(\varepsilon)}, i,j \in \mathbb{X}$. Thus, $E_{ii}(1,\varepsilon) = \sum_{j\in\mathbb{X}} e_j(\varepsilon)\frac{\rho_j(\varepsilon)}{\rho_i(\varepsilon)}, i \in \mathbb{X}$. By substituting this expressions for expectations $E_{ii}(1,\varepsilon)$ into relation (3.18), we get the following useful formula, which connect stationary distributions of the embedded Markov chain $\eta_{\varepsilon,n}$ and the semi-Markov process $\eta_\varepsilon(t)$,

$$\pi_i(\varepsilon) = \frac{e_i(\varepsilon)\rho_i(\varepsilon)}{\sum_{j\in\mathbb{X}} e_j(\varepsilon)\rho_j(\varepsilon)}, \ i \in \mathbb{X}. \tag{3.20}$$

This relation implies that $\rho_i(\varepsilon) = \pi_i(\varepsilon)e_i^{-1}(\varepsilon)e(\varepsilon), i \in \mathbb{X}$, where factor $e(\varepsilon) = \sum_{j\in\mathbb{X}} e_j(\varepsilon)\rho_j(\varepsilon)$. By substituting these expressions in system (3.16) and then canceling factor $e(\varepsilon)$ in their equations, we get the system of linear equations (3.19). Since system (3.16) has the unique solution, system (3.19) also has the unique solution.

3.1.4 Perturbation Conditions for Semi-Markov Processes

Let us assume that the following perturbation condition, based on Taylor asymptotic expansions, holds:

D: $p_{ij}(\varepsilon) = \sum_{l=l_{ij}^-}^{l_{ij}^+} a_{ij}[l]\varepsilon^l + o_{ij}(\varepsilon^{l_{ij}^+})$, $\varepsilon \in (0, \varepsilon_0]$, for $j \in \mathbb{Y}_i, i \in \mathbb{X}$, where **(a)** coefficients $a_{ij}[k], l_{ij}^- \leq k \leq l_{ij}^+$ are real numbers, $a_{ij}[l_{ij}^-] > 0$ and $0 \leq l_{ij}^- \leq l_{ij}^+ < \infty$ are integers, for $j \in \mathbb{Y}_i, i \in \mathbb{X}$; **(b)** function $o_{ij}(\varepsilon^{l_{ij}^+})/\varepsilon^{l_{ij}^+} \to 0$ as $\varepsilon \to 0$, for $j \in \mathbb{Y}_i, i \in \mathbb{X}$.

We also assume that the following perturbation condition, based on Laurent asymptotic expansions, holds:

$\mathbf{E}_d$: $e_{ij}(k,\varepsilon) = \sum_{l=m_{ij}^-[k]}^{m_{ij}^+[k]} b_{ij}[k,l]\varepsilon^l + \dot{o}_{k,ij}(\varepsilon^{m_{ij}^+[k]})$, $\varepsilon \in (0, \varepsilon_0]$, for $k = 1,\dots,d, j \in \mathbb{Y}_i, i \in \mathbb{X}$, where **(a)** coefficients $b_{ij}[k,l], m_{ij}^-[k] \leq l \leq m_{ij}^+[k]$ are real numbers, $b_{ij}[k, m_{ij}^-[k]] > 0$ and $-\infty < m_{ij}^-[k] \leq m_{ij}^+[k] < \infty$ are integers, for $k = 1,\dots,d, j \in \mathbb{Y}_i, i \in \mathbb{X}$; **(b)** function $\dot{o}_{k,ij}(\varepsilon^{m_{ij}^+[k]})/\varepsilon^{m_{ij}^+[k]} \to 0$ as $\varepsilon \to 0$, for $k = 1,\dots,d, j \in \mathbb{Y}_i, i \in \mathbb{X}$.

According to the remarks made in Section 2.1, the perturbation type of asymptotic expansions appearing in conditions **D** and $\mathbf{E}_d$ can be classified via their lengths. Respectively, it is natural to classify the type of perturbation for the semi-Markov process $\eta_\varepsilon(t)$ via the maximal length of asymptotic expansions in the above conditions, $L_{\max} = \max_{j\in\mathbb{Y}_i, i\in\mathbb{X}} \left((l_{ij}^+ - l_{ij}^-) \vee \max_{1\leq k\leq d}(m_{ij}^+[k] - m_{ij}^-[k])\right)$. The perturbation is of linear or nonlinear type if, respectively, $L_{\max} = 1$ or $L_{\max} > 1$.

If $\eta^{(\varepsilon)}(t)$ is a discrete time Markov chain, condition **D** implies condition $\mathbf{E}_d$, since, in this case, expectations $e_{ij}(k,\varepsilon) = p_{ij}(\varepsilon), k = 1,\dots,d, j \in \mathbb{Y}_i, i \in \mathbb{X}$.

If $\eta^{(\varepsilon)}(t)$ is a continuous time Markov chain, condition $\mathbf{E}_d$ can be replaced by an analogous condition, which assumes that expectations $e_i(\varepsilon) = \lambda_i(\varepsilon)^{-1}, i \in \mathbb{X}$ can be represented in the form of pivotal Laurent asymptotic expansions. This condition and condition **D** would imply condition $\mathbf{E}_d$ to hold, with the corresponding Laurent asymptotic expansions obtained by application proposition **(ii)** (the multiple multiplication rule) of Lemma 2.3 to the products $e_{ij}(k, \varepsilon) = k! e_i^k(\varepsilon) p_{ij}(\varepsilon), k = 1, \ldots, d, j \in \mathbb{Y}_i, i \in \mathbb{X}$.

Conditions **A**, **D**, and $\mathbf{E}_d$, assumed to hold for some $\varepsilon_0 \in (0,1]$, also hold for any $\varepsilon'_0 \in (0, \varepsilon_0]$.

It worth to note that an actual value of parameter $\varepsilon_0 \in (0,1]$ is not important in propositions concerned asymptotic expansions with remainders given in the standard form of $o(\cdot)$.

Let us, for the moment, exclude sub-condition **(a)** from condition **A**. Conditions **D** and $\mathbf{E}_d$ imply that there exits $\tilde{\varepsilon}_0 \in (0, \varepsilon_0]$ such that $p_{ij}(\varepsilon) = \sum_{l=l_{ij}^-}^{l_{ij}^+} a_{ij}[l]\varepsilon^l + o_{ij}(\varepsilon^{l_{ij}^+}) > 0$, for $j \in \mathbb{Y}_i$, $i \in \mathbb{X}$, $\varepsilon \in (0, \tilde{\varepsilon}_0]$ and $e_{ij}(k, \varepsilon) = \sum_{l=m_{ij}^-[k]}^{m_{ij}^+[k]} b_{k,ij}[l]\varepsilon^l + \dot{o}_{k,ij}(\varepsilon^{m_{ij}^+[k]}) > 0$, for $k = 1, \ldots, d, j \in \mathbb{Y}_i, i \in \mathbb{X}, \varepsilon \in (0, \tilde{\varepsilon}_0]$. We can, just, decrease parameter ε_0 and take the new $\varepsilon_0 = \tilde{\varepsilon}_0$. Condition **A (a)** holds for this new value of ε_0.

We, however, do prefer to include sub-condition **(a)** in condition **A**, in order to have a clear description for the communicative structure of the phase space $\mathbb{X}$, in one condition. In this case, the above inequalities hold for $\tilde{\varepsilon}_0 = \varepsilon_0$.

Conditions **D** and $\mathbf{E}_d$ are consistent with condition **A (a)**.

Matrix $\| p_{ij}(\varepsilon) \|$ is stochastic, for every $\varepsilon \in (0, \varepsilon_0]$. This model stochasticity assumption holds by the default.

Condition **D** should, also, be consistent with this model stochasticity assumption.

Condition **D** and proposition **(i)** (the multiple summation rule) of Lemma 2.3 imply that sum $\sum_{j \in \mathbb{Y}} p_{ij}(\varepsilon)$ can, for every subset $\mathbb{Y} \subseteq \mathbb{Y}_i$ and $i \in \mathbb{X}$, be represented in the form of the following Taylor asymptotic expansion:

$$\sum_{j \in \mathbb{Y}} p_{ij}(\varepsilon) = \sum_{l=l_{i,\mathbb{Y}}^-}^{l_{i,\mathbb{Y}}^+} a_{i,\mathbb{Y}}[l]\varepsilon^l + o_{i,\mathbb{Y}}(\varepsilon^{l_{i,\mathbb{Y}}^+}), \tag{3.21}$$

where: (a) $l_{i,\mathbb{Y}}^{\pm} = \min_{j \in \mathbb{Y}} l_{ij}^{\pm}$, (b) $a_{i,\mathbb{Y}}[l] = \sum_{j \in \mathbb{Y}} a_{ij}[l]$, $l = l_{i,\mathbb{Y}}^-, \ldots, l_{i,\mathbb{Y}}^+$, where $a_{ij}[l] = 0$, for $0 \leq l < l_{ij}^-, j \in \mathbb{Y}$, and (c) $o_{i,\mathbb{Y}}(\varepsilon^{l_{i,\mathbb{Y}}^+}) = \sum_{j \in \mathbb{Y}} (\sum_{l_{i,\mathbb{Y}}^+ < l \leq l_{ij}^+} a_{ij}[l]\varepsilon^l + o_{ij}(\varepsilon^{l_{ij}^+}))$.

Let us introduce the following condition, which reflects additional links (which are caused by the above model stochasticity assumption) between the asymptotic expansions appearing in condition **D**:

F: **(a)** $a_{i,\mathbb{Y}_i}[l] = \sum_{j \in \mathbb{Y}_i} a_{ij}[l] = \mathrm{I}(l = 0)$, $0 = l_{i,\mathbb{Y}_i}^- \leq l \leq l_{i,\mathbb{Y}_i}^+$, $i \in \mathbb{X}$, where $a_{ij}[l] = 0$, for $0 \leq l < l_{ij}^-, j \in \mathbb{Y}_i, i \in \mathbb{X}$; **(b)** $o_{i,\mathbb{Y}_i}(\varepsilon^{l_{i,\mathbb{Y}_i}^+}) = 0, \varepsilon \in (0, \varepsilon_0], i \in \mathbb{X}$.

Lemma 3.1. *Let conditions* **A (a)–(b)** *and* **D** *hold. In this case, condition* **F** *is equivalent to the model stochasticity assumption that matrix* $\|p_{ij}(\varepsilon)\|$ *is stochastic, for every* $\varepsilon \in (0, \varepsilon_0]$.

Proof. The model stochasticity assumption for matrices $\|p_{ij}(\varepsilon)\|$, $\varepsilon \in (0, \varepsilon_0]$, takes, under conditions **A (a)–(b)**, the form of the following identity, which should hold for every $i \in \mathbb{X}$:

$$\sum_{j \in \mathbb{Y}_i} p_{ij}(\varepsilon) = 1, \ \varepsilon \in (0, \varepsilon_0]. \tag{3.22}$$

Condition **D** and Lemma 2.3 let us write down the asymptotic expansion (2.20) for the case $\mathbb{Y} = \mathbb{Y}_i$. Constant 1 also can be interpreted as the asymptotic expansion $1 = 1 + 0\varepsilon + \cdots + 0\varepsilon^k + o(\varepsilon^k)$ for $k = l_{i,Y_i}$ and $o(\varepsilon^k) \equiv 0$. Then, identity (3.22) let one apply Lemma 2.1 to the described above two asymptotic expansions and get relations appearing in condition **F**. In Lemma 2.1, the assumptions are made that conditions **A (a)–(b)** and **D** hold. Conditions **A (a)–(b)** imply that $p_{ij}(\varepsilon) \geq 0$ for $i, j \in \mathbb{X}$. In this case, conditions **D** and **F** obviously imply that $\sum_{j \in \mathbb{X}} p_{ij}(\varepsilon) = 1$, for $i \in \mathbb{X}$. Thus, matrix $\|p_{ij}(\varepsilon)\|$ is stochastic. □

Also, some additional conditions should hold for the Laurent asymptotic expansions for transition probabilities and moments of transition times, which appear in conditions **D** and $\mathbf{E}_d$. These conditions are caused by Stieltjes' moment inequalities (which appear in the Stieltjes' moment problem) connecting power moments $e_{ij}(k, \varepsilon) = \int_0^\infty t^k Q_{\varepsilon,ij}(dt), k = 0, \ldots, d, j \in \mathbb{Y}_i, i \in \mathbb{X}$.

Let us introduce matrices $\mathbf{E}_{i,j}(k, \varepsilon), \varepsilon \in (0, \varepsilon_0], k = 0, 1, \ldots, d, \ j \in \mathbb{Y}_i, \ i \in \mathbb{X}$,

$$\mathbf{E}_{ij}(k, \varepsilon) = \begin{cases} \|e_{ij}(k' + k'', \varepsilon)\|_{k',k''=0}^n & \text{if } k = 2n \text{ is even number,} \\ \|e_{ij}(k' + k'' + 1, \varepsilon)\|_{k',k''=0}^n & \text{if } k = 2n + 1 \text{ is odd number.} \end{cases} \tag{3.23}$$

Let us now define the determinant functions, $\Delta_{ij}(k, \varepsilon), \varepsilon \in (0, \varepsilon_0]$, for $k = 0, \ldots, d$, $j \in \mathbb{Y}_i, i \in \mathbb{X}$,

$$\Delta_{ij}(k, \varepsilon) = \det(\mathbf{E}_{ij}(k, \varepsilon)). \tag{3.24}$$

Obviously, $\Delta_{ij}(0, \varepsilon) = e_{ij}(0, \varepsilon), \Delta_{ij}(1, \varepsilon) = e_{ij}(1, \varepsilon), \Delta_{ij}(2, \varepsilon) = e_{ij}(2, \varepsilon) e_{ij}(0, \varepsilon) - e_{ij}^2(1, \varepsilon), \ldots$, for $j \in \mathbb{Y}_i, i \in \mathbb{X}$.

According to condition $\mathbf{C}_d$, moments $e_{ij}(k, \varepsilon), \varepsilon \in (0, \varepsilon_0], k = 0, \ldots, d, \ j \in \mathbb{Y}_i, i \in \mathbb{X}$ are finite and, therefore, the Stieltjes' moment inequalities imply that, the following condition should hold:

$\mathbf{G}_d$: $\Delta_{ij}(k, \varepsilon) = \det(\mathbf{E}_{ij}(k, \varepsilon)) \geq 0, \ \varepsilon \in (0, \varepsilon_0], \ k = 0, \ldots, d, j \in \mathbb{Y}_i, i \in \mathbb{X}$.

By applying Lemmas 2.1–2.8 to functions $\Delta_{ij}(k, \varepsilon), \varepsilon \in (0, \varepsilon_0]$, one can, for every $k = 0, \ldots, d, j \in \mathbb{Y}_i, i \in \mathbb{X}$ represent these functions in the form of Laurent asymptotic expansions,

$$\Delta_{ij}(k, \varepsilon) = \sum_{l = L_{ij}^-[k]}^{L_{ij}^+[k]} \delta_{ij}[k, l] \varepsilon^l + \dot{o}_{k,ij}(\varepsilon^{L_{ij}^+[k]}), \varepsilon \in (0, \varepsilon_0]. \tag{3.25}$$

The above expansions may not be pivotal, since determinants $\Delta_{ij}(k,\varepsilon)$ are some of products of moments with both signs $+$ and $-$. In particular, it is possible that all coefficients in sequence $\delta_{ij}[k,l], l = L^-_{ij}[k], \ldots, L^+_{ij}[k]$ are zeros.

However, if there exist non-zero terms in this sequence, then the first such term should take a positive value. Otherwise, the function $\Delta_{ij}(k,\varepsilon)$ would be negative for ε small enough.

Conditions **A** and **D** imply that there exist limits,

$$p_{ij}(0) = \lim_{\varepsilon\to 0} p_{ij}(\varepsilon) = \begin{cases} a_{ij}[0] & \text{if } l^-_{ij} = 0, j \in \mathbb{Y}_i, i \in \mathbb{X}, \\ 0 & \text{if } l^-_{ij} > 0, j \in \mathbb{Y}_i, i \in \mathbb{X}, \text{ or } j \in \overline{\mathbb{Y}}_i, i \in \mathbb{X}. \end{cases} \tag{3.26}$$

Matrix $\|p_{ij}(\varepsilon)\|$ is stochastic, for every $\varepsilon \in (0,\varepsilon_0]$, and, thus, matrix $\|p_{ij}(0)\|$ is also stochastic. However, it is possible that matrix $\|p_{ij}(0)\|$ has more zero elements than matrices $\|p_{ij}(\varepsilon)\|$. Therefore, the Markov chain $\eta_{0,n}$, with the phase space $\mathbb{X}$ and the matrix of transition probabilities $\|p_{ij}(0)\|$ can be non-ergodic, and its phase space $\mathbb{X}$ can consist of one or several closed classes of communicative states plus, possibly, a class of transient states.

Conditions **A**, $\mathbf{C}_d$, and $\mathbf{E}_d$ imply that there exist limits, for $k = 1,\ldots,d$,

$$e_{ij}(k,0) = \lim_{\varepsilon\to 0} e_{ij}(k,\varepsilon) = \begin{cases} \infty & \text{if } m^-_{ij}[k] < 0, j \in \mathbb{Y}_i, i \in \mathbb{X}, \\ b_{ij}[k,0] & \text{if } m^-_{ij}[k] = 0, j \in \mathbb{Y}_i, i \in \mathbb{X}, \\ 0 & \text{if } m^-_{ij}[k] > 0, j \in \mathbb{Y}_i, i \in \mathbb{X}, \\ 0 & \text{or } j \in \overline{\mathbb{Y}}_i, i \in \mathbb{X}. \end{cases} \tag{3.27}$$

Our goal is to compose effective algorithms for construction of asymptotic expansions for moments of hitting times $E_{ij}(k,\varepsilon), k = 1,\ldots,i, j \in \mathbb{X}$, and for stationary probabilities $\pi_i(\varepsilon), i \in \mathbb{X}$.

As we shall see, the proposed algorithms can be applied to models with an arbitrary asymptotic communicative structure of phase spaces and all types of asymptotic behavior for moments of sojourn times given in relation (3.27).

If probabilities $p_{ij}(\varepsilon)$ and moments $e_{ij}(k,\varepsilon), k = 1,\ldots,d$ are given as initial characteristics, then conditions **D** and $\mathbf{E}_d$, are, in fact, equivalent to the assumption of existence the derivative-like limits at zero, $\lim_{\varepsilon\to 0} \frac{\varepsilon^{-l^-_{ij}} p_{ij}(\varepsilon) - \sum_{l=0}^{r-1} a_{ij}[l^-_{ij}+l]\varepsilon^l}{\varepsilon^r} = a_{ij}[l^-_{ij} + r], r = 0,\ldots l^+_{ij} - l^-_{ij}$ and $\lim_{\varepsilon\to 0} \frac{\varepsilon^{-m^-_{ij}[k]} e_{ij}(k,\varepsilon) - \sum_{l=0}^{r-1} b_{ij}[k, m^-_{ij}[k]+l]\varepsilon^l}{\varepsilon^r} = b_{ij}[m^-_{ij}[k] + r], r = 0,\ldots m^+_{ij}[k] - m^-_{ij}[k], k = 1,\ldots,d$, for $j \in \mathbb{Y}_i, i \in \mathbb{X}$. If these limits are computed correctly, then conditions **F** and $\mathbf{G}_d$ hold automatically. Moreover, these conditions do not affect formulas for computing parameters and coefficients of Laurent asymptotic expansions for power moment of hitting times.

We are also interested in asymptotic expansions with explicit upper bounds for remainders.

In this case, the perturbation condition **D** should be replaced by the following stronger condition, in which the corresponding Taylor asymptotic expansions are given in the form with explicit upper bounds for remainders:

D$'$: $p_{ij}(\varepsilon) = \sum_{l=l_{ij}^-}^{l_{ij}^+} a_{ij}[l]\varepsilon^l + o_{ij}(\varepsilon^{l_{ij}^+}), \varepsilon \in (0, \varepsilon_0]$, for $j \in \mathbb{Y}_i, i \in \mathbb{X}$, where **(a)** coefficients $a_{ij}[k], l_{ij}^- \leq k \leq l_{ij}^+$ are real numbers, $a_{ij}[l_{ij}^-] > 0$ and $0 \leq l_{ij}^- \leq l_{ij}^+ < \infty$ are integers, for $j \in Y_i, i \in X$; **(b)** $|o_{ij}(\varepsilon^{l_{ij}^+})| \leq G_{ij}\varepsilon^{l_{ij}^+ + \delta_{ij}}, 0 < \varepsilon \leq \varepsilon_{ij}$, for $j \in \mathbb{Y}_i, i \in \mathbb{X}$, where $0 < \delta_{ij} \leq 1$, $0 \leq G_{ij} < \infty$, $0 < \varepsilon_{ij} \leq \varepsilon_0$.

Also, the perturbation condition $\mathbf{E}_d$ should be replaced by the following stronger condition, in which the corresponding Laurent asymptotic expansions are given in the form with explicit upper bounds for remainders:

$\mathbf{E}_d'$: $e_{ij}(k, \varepsilon) = \sum_{l=m_{ij}^-[k]}^{m_{ij}^+[k]} b_{ij}[k, l]\varepsilon^l + \dot{o}_{k,ij}(\varepsilon^{m_{ij}^+}), \varepsilon \in (0, \varepsilon_0]$, for $k = 1, \ldots, d$, $j \in \mathbb{Y}_i$, $i \in \mathbb{X}$, where **(a)** coefficients $b_{ij}[k, l], m_{ij}^-[k] \leq l \leq m_{ij}^+[k]$ are real numbers, $b_{ij}[k, m_{ij}^-[k]] > 0$ and $-\infty < m_{ij}^-[k] \leq m_{ij}^+[k] < \infty$ are integers, for $k = 1, \ldots, d, j \in \mathbb{Y}_i, i \in \mathbb{X}$; **(b)** $|\dot{o}_{k,ij}(\varepsilon^{m_{ij}^+}[k])| \leq \dot{G}_{ij}[k]\varepsilon^{m_{ij}^+[k] + \dot{\delta}_{ij}[k]}, 0 < \varepsilon \leq \dot{\varepsilon}_{ij}[k]$, for $k = 1, \ldots, d, j \in \mathbb{Y}_i, i \in \mathbb{X}$, where $0 < \dot{\delta}_{ij}[k] \leq 1$, $0 \leq \dot{G}_{ij}[k] < \infty$, $0 < \dot{\varepsilon}_{ij}[k] \leq \varepsilon_0$, for $k = 1, \ldots, d, j \in \mathbb{Y}_i, i \in \mathbb{X}$.

Conditions **D**$'$ and $\mathbf{E}_d'$ differ of conditions **D** and $\mathbf{E}_d$ by the assumptions imposed on the remainders of the asymptotic expansions appearing in these conditions. In conditions **D** and $\mathbf{E}_d$, the remainders are given in the standard form of $o(\cdot)$. In conditions **D**$'$ and $\mathbf{E}_d'$ the remainders are given in the form with explicit upper bounds.

We use periods above the letters denoting parameters and remainders of asymptotic expansions, which appear in conditions $\mathbf{E}_d$ and $\mathbf{E}_d'$, in order to distinguish these parameters and remainders of asymptotic expansions, which appear in conditions **D** and **D**$'$.

Since conditions **D**$'$ and $\mathbf{E}_d'$ are stronger than, respectively, conditions **D** and $\mathbf{E}_d$, conditions **F** and $\mathbf{G}_d$ also hold.

It is also worth to note that the models of nonlinearly perturbed discrete and continuous Markov chains are particular cases of the above model of nonlinearly perturbed semi-Markov processes.

3.2 Reduction of Phase Spaces for Semi-Markov Processes

In Section 3.2, we describe a one-step time-space screening procedure of phase space reduction for perturbed semi-Markov processes, introduce hitting times, and prove invariant property for moments of hitting times with respect to the procedure of phase space reduction.

3.2.1 Semi-Markov Processes with Reduced Phase Spaces

Let us choose some state $r \in \mathbb{X}$ and consider the reduced phase space ${}_r\mathbb{X} = \mathbb{X} \setminus \{r\}$, with the state r excluded from the phase space $\mathbb{X}$.

Let us define the sequential moments of hitting the reduced space ${}_r\mathbb{X}$ by the embedded Markov chain $\eta_{\varepsilon,n}$,

$$ {}_r\xi_{\varepsilon,n} = \min(l > {}_r\xi_{\varepsilon,n-1},\ \eta_{\varepsilon,l} \in {}_r\mathbb{X}),\ n = 1,2,\ldots,\ {}_r\xi_{\varepsilon,0} = 0. \quad (3.28) $$

Now, let us define the random sequence,

$$ ({}_r\eta_{\varepsilon,n}, {}_r\kappa_{\varepsilon,n}) = \begin{cases} (\eta_{\varepsilon,0}, 0) & \text{for } n = 0, \\ (\eta_{\varepsilon,{}_r\xi_{\varepsilon,n}}, \sum_{l={}_r\xi_{\varepsilon,n-1}+1}^{{}_r\xi_{\varepsilon,n}} \kappa_{\varepsilon,l}) & \text{for } n = 1,2,\ldots. \end{cases} \quad (3.29) $$

This sequence is also a Markov renewal process with a phase space $\mathbb{X} \times [0,\infty)$, the initial distribution $\bar{p}_\varepsilon = \langle p_{\varepsilon,i} = \mathsf{P}\{{}_r\eta_{\varepsilon,0} = i,\ {}_r\kappa_{\varepsilon,0} = 0\} = \mathsf{P}\{{}_r\eta_{\varepsilon,0} = i\}, i \in \mathbb{X}\rangle$, and transition probabilities defined for $(i,s),(j,t) \in \mathbb{X} \times [0,\infty)$,

$$ \begin{aligned} {}_rQ_{\varepsilon,ij}(t) &= \mathsf{P}\{{}_r\eta_{\varepsilon,1} = j,\ {}_r\kappa_{\varepsilon,1} \le t / {}_r\eta_{\varepsilon,0} = i,\ {}_r\kappa_{\varepsilon,0} = s\} \\ &= Q_{\varepsilon,ij}(t) + \sum_{n=0}^{\infty} Q_{\varepsilon,ir}(t) * Q^{*n}_{\varepsilon,rr}(t) * Q_{\varepsilon,rj}(t). \end{aligned} \quad (3.30) $$

Here, symbol $*$ is used to denote the convolution of distribution functions (possibly improper), and $Q^{*n}_{\varepsilon,rr}(t)$ is the n times convolution of the distribution function $Q_{\varepsilon,rr}(t)$.

The corresponding embedded Markov chain ${}_r\eta_{\varepsilon,n}$ has the phase space $\mathbb{X}$, the initial distribution $\bar{p}_\varepsilon = \langle p_{\varepsilon,i} = \mathsf{P}\{{}_r\eta_{\varepsilon,0} = i\}, i \in \mathbb{X}\rangle$, and transition probabilities defined for $i,j \in \mathbb{X}$,

$$ \begin{aligned} {}_rp_{ij}(\varepsilon) &= p_{ij}(\varepsilon) + \sum_{n=0}^{\infty} p_{ir}(\varepsilon) p_{rr}(\varepsilon)^n p_{rj}(\varepsilon) \\ &= \begin{cases} \frac{p_{rj}(\varepsilon)}{1-p_{rr}(\varepsilon)} & \text{for } i = r, \\ p_{ij}(\varepsilon) + p_{ir}(\varepsilon)\frac{p_{rj}(\varepsilon)}{1-p_{rr}(\varepsilon)} & \text{for } i \in {}_r\mathbb{X}. \end{cases} \end{aligned} \quad (3.31) $$

It is useful to note that the second formula in relation (3.31) reduces to the first one, if to assign $i = r$ in this formula.

Condition **A** implies that probabilities $p_{rr}(\varepsilon) \in [0,1)$, for $r \in \mathbb{X}$, $\varepsilon \in (0,\varepsilon_0]$.

The transition distributions for the Markov chain ${}_r\eta_{\varepsilon,n}$, are concentrated on the reduced phase space ${}_r\mathbb{X}$, i.e., for every $i \in \mathbb{X}$,

$$\sum_{j\in {}_r\mathbb{X}} {}_rp_{ij}(\varepsilon) = \sum_{j\in {}_r\mathbb{X}} p_{ij}(\varepsilon) + p_{ir}(\varepsilon) \sum_{j\in {}_r\mathbb{X}} \frac{p_{rj}(\varepsilon)}{1-p_{rr}(\varepsilon)}$$
$$= \sum_{j\in {}_r\mathbb{X}} p_{ij}(\varepsilon) + p_{ir}(\varepsilon) = 1. \tag{3.32}$$

If the initial distribution $\bar{p}_\varepsilon$ is concentrated on the phase space ${}_r\mathbb{X}$, i.e., $p_{\varepsilon,r} = 0$, then the random sequence $({}_r\eta_{\varepsilon,n}, {}_r\kappa_{\varepsilon,n}), n = 0,1,\dots$ is a Markov renewal process with the reduced phase ${}_r\mathbb{X}\times[0,\infty)$, the initial distribution ${}_r\bar{p}_\varepsilon = \langle p_{\varepsilon,i} = \mathsf{P}\{{}_r\eta_{\varepsilon,0} = i, {}_r\kappa_{\varepsilon,0} = 0\} = \mathsf{P}\{{}_r\eta_{\varepsilon,0} = i\}, i \in {}_r\mathbb{X}\rangle$, and transition probabilities ${}_rQ_{\varepsilon,ij}(t), t \geq 0, i,j \in {}_r\mathbb{X}$.

If the initial distribution $\bar{p}$ is not concentrated on the phase space ${}_r\mathbb{X}$, i.e., $p_{\varepsilon,r} > 0$, then the random sequence $({}_r\eta_{\varepsilon,n}, {}_r\kappa_{\varepsilon,n}), n = 0,1,\dots$ can be interpreted as a Markov renewal process with the so-called transition period.

Respectively, one can define the transformed semi-Markov process with the reduced phase space ${}_r\mathbb{X}$,

$${}_r\eta_\varepsilon(t) = {}_r\eta_{\varepsilon,{}_r\nu_\varepsilon(t)},\ t \geq 0, \tag{3.33}$$

where ${}_r\nu_\varepsilon(t) = \max(n \geq 0 : {}_r\zeta_{\varepsilon,n} \leq t)$ is a number of jumps at time interval $[0,t]$, for $t \geq 0$, and ${}_r\zeta_{\varepsilon,n} = {}_r\kappa_{\varepsilon,1} + \cdots + {}_r\kappa_{\varepsilon,n},\ n = 0,1,\dots$ are sequential moments of jumps, for the semi-Markov process ${}_r\eta_\varepsilon(t)$.

If the initial distribution $\bar{p}_\varepsilon$ is concentrated on the phase space ${}_r\mathbb{X}$, then process ${}_r\eta_\varepsilon(t)$ is a standard semi-Markov process with the reduced phase ${}_r\mathbb{X}$, the initial distribution ${}_r\bar{p}_\varepsilon = \langle {}_rp_i = \mathsf{P}\{{}_r\eta_\varepsilon(0) = i\}, i \in {}_r\mathbb{X}\rangle$, and transition probabilities ${}_rQ_{\varepsilon,ij}(t), t \geq 0, i,j \in {}_r\mathbb{X}$.

According to the above remarks, we can refer to the process ${}_r\eta_\varepsilon(t)$ as a reduced semi-Markov process.

If the initial distribution $\bar{p}_\varepsilon$ is not concentrated on the phase space ${}_r\mathbb{X}$, then the process ${}_r\eta_\varepsilon(t)$ can be interpreted as a reduced semi-Markov process with transition period.

Let us introduce moment of transition times for the reduced semi-Markov processes ${}_r\eta_\varepsilon(t)$,

$${}_re_{ij}(k,\varepsilon) = \int_0^\infty t\ {}_rQ_{\varepsilon,ij}(dt),\ k = 0,\dots,d,\ i,j \in {}_r\mathbb{X}. \tag{3.34}$$

Conditions **A** and $\mathbf{C}_d$ readily imply (see, for example, Silvestrov and Manca (2017)) that, for all $r, i \in \mathbb{X}, j \in {}_r\mathbb{X}$ and $\varepsilon \in (0,\varepsilon_0]$,

$${}_re_{ij}(d,\varepsilon) < \infty. \tag{3.35}$$

Note that, ${}_re_{ij}(0,\varepsilon) = {}_rp_{ij}(\varepsilon),\ i \in \mathbb{X}, j \in {}_r\mathbb{X}$.

Let now get explicit recurrent formulas for high order moments, ${}_re_{ij}(k,\varepsilon), k = 1,\dots,d,\ i,j \in \mathbb{X}$, for the reduced semi-Markov process ${}_r\eta_\varepsilon(t)$.

Taking into account that ${}_r\xi_{\varepsilon,1}$ is Markov time for the Markov renewal process $({}_r\eta_{\varepsilon,n}, {}_r\kappa_{\varepsilon,n})$, we can write down the following system of stochastic equalities, for every $i,j \in {}_r\mathbb{X}$:

$$\begin{cases} {}_r\kappa_{\varepsilon,r,1}\mathrm{I}({}_r\eta_{\varepsilon,r,1} = j) \stackrel{d}{=} \kappa_{\varepsilon,r,1}\mathrm{I}(\eta_{\varepsilon,r,1} = j) \\ \qquad\qquad + (\kappa_{\varepsilon,r,1} + {}_r\kappa_{\varepsilon,r,1})\mathrm{I}(\eta_{\varepsilon,r,1} = r)\mathrm{I}({}_r\eta_{\varepsilon,r,1} = j), \\ {}_r\kappa_{\varepsilon,i,1}\mathrm{I}({}_r\eta_{\varepsilon,i,1} = j) \stackrel{d}{=} \kappa_{\varepsilon,i,1}\mathrm{I}(\eta_{\varepsilon,i,1} = j) \\ \qquad\qquad + (\kappa_{\varepsilon,i,1} + {}_r\kappa_{\varepsilon,r,1})\mathrm{I}(\eta_{\varepsilon,i,1} = r)\mathrm{I}({}_r\eta_{\varepsilon,r,1} = j), \end{cases} \tag{3.36}$$

where (a) $(\eta_{\varepsilon,i,1}, \kappa_{\varepsilon,i,1})$ is a random vector, which takes values in space $\mathbb{X} \times [0,\infty)$ and has the distribution $\mathsf{P}\{\eta_{\varepsilon,i,1} = j, \kappa_{\varepsilon,i,1} \leq t\} = \mathsf{P}_i\{\eta_{\varepsilon,1} = j, \kappa_{\varepsilon,1} \leq t\} = Q_{\varepsilon,ij}(t)$, $j \in \mathbb{X}, t \geq 0$, for every $i \in \mathbb{X}$; (b) $({}_r\eta_{\varepsilon,i,1}, {}_r\kappa_{\varepsilon,i,1})$ is a random vector which takes values in the space ${}_r\mathbb{X} \times [0,\infty)$ and has distribution $\mathsf{P}\{{}_r\eta_{\varepsilon,i,1} = j, {}_r\kappa_{\varepsilon,i,1} \leq t\} = \mathsf{P}_i\{{}_r\eta_{\varepsilon,1} = j, {}_r\kappa_{\varepsilon,1} \leq t\} = {}_rQ_{\varepsilon,ij}(t)$, $j \in {}_r\mathbb{X}, t \geq 0$, for every $i \in \mathbb{X}$; (c) $(\eta_{\varepsilon,i,1}, \kappa_{\varepsilon,i,1})$ and $({}_r\eta_{\varepsilon,r,1}, {}_r\kappa_{\varepsilon,r,1})$ are independent random vectors, for every $i, r \in \mathbb{X}$.

By taking moments of the order k in stochastic relations (3.36) we get, for every $i, j \in {}_r\mathbb{X}, r \in \mathbb{X}$, the following system of linear equations for moments ${}_re_{rj}(k,\varepsilon)$, ${}_re_{ij}(k,\varepsilon)$, for $k = 1,\dots,d$,

$$\begin{cases} {}_re_{rj}(k,\varepsilon) = e_{rj}(k,\varepsilon) + \sum_{l=0}^{k-1}\binom{k}{l} e_{rr}(k-l,\varepsilon)\,{}_re_{rj}(l,\varepsilon) + p_{rr}(\varepsilon)\,{}_re_{rj}(k,\varepsilon), \\ {}_re_{ij}(k,\varepsilon) = e_{ij}(k,\varepsilon) + \sum_{l=0}^{k-1}\binom{k}{l} e_{ir}(k-l,\varepsilon)\,{}_re_{rj}(l,\varepsilon) + p_{ir}(\varepsilon)\,{}_re_{rj}(k,\varepsilon). \end{cases} \tag{3.37}$$

Relation (3.37) yields, for every $i, j \in {}_r\mathbb{X}, r \in \mathbb{X}$, the following recurrent formulas for moments ${}_re_{rj}(k,\varepsilon)$, ${}_re_{ij}(k,\varepsilon)$, which should be used recurrently, for $k = 1,\dots,d$:

$$\begin{cases} {}_re_{rj}(k,\varepsilon) = \frac{1}{1-p_{rr}(\varepsilon)}\big(e_{rj}(k,\varepsilon) + \sum_{l=0}^{k-1}\binom{k}{l} e_{rr}(k-l,\varepsilon)\,{}_re_{rj}(l,\varepsilon)\big), \\ {}_re_{ij}(k,\varepsilon) = e_{ij}(k,\varepsilon) + \sum_{l=0}^{k-1}\binom{k}{l} e_{ir}(k-l,\varepsilon)\,{}_re_{rj}(l,\varepsilon) + p_{ir}(\varepsilon)\,{}_re_{rj}(k,\varepsilon) \\ \qquad = e_{ij}(k,\varepsilon) + \sum_{l=0}^{k-1}\binom{k}{l} e_{ir}(k-l,\varepsilon)\,{}_re_{rj}(l,\varepsilon) \\ \qquad\quad + \frac{p_{ir}(\varepsilon)}{1-p_{rr}(\varepsilon)}\big(e_{rj}(k,\varepsilon) + \sum_{l=0}^{k-1}\binom{k}{l} e_{rr}(k-l,\varepsilon)\,{}_re_{rj}(l,\varepsilon)\big). \end{cases} \tag{3.38}$$

It is useful to note that the second formula in relation (3.38) reduces to the first one, if to assign $i = r$ in this formula.

In particular, relation (3.38) implies the following formulas for first two moments, for every $i, j \in {}_r\mathbb{X}, r \in \mathbb{X}$:

$$\begin{cases} {}_re_{rj}(1,\varepsilon) = \frac{1}{1-p_{rr}(\varepsilon)}\big(e_{rj}(1,\varepsilon) + e_{rr}(1,\varepsilon)\,{}_re_{rj}(0,\varepsilon)\big), \\ \qquad = \frac{1}{1-p_{rr}(\varepsilon)}\big(e_{rj}(1,\varepsilon) + e_{rr}(1,\varepsilon)\,\frac{p_{rj}(\varepsilon)}{1-p_{rr}(\varepsilon)}\big), \\ {}_re_{ij}(1,\varepsilon) = e_{ij}(1,\varepsilon) + e_{ir}(1,\varepsilon)\,{}_re_{rj}(0,\varepsilon) + p_{ir}(\varepsilon)\,{}_re_{rj}(1,\varepsilon) \\ \qquad = e_{ij}(1,\varepsilon) + e_{ir}(1,\varepsilon)\,\frac{p_{rj}(\varepsilon)}{1-p_{rr}(\varepsilon)} \\ \qquad\quad + \frac{p_{ir}(\varepsilon)}{1-p_{rr}(\varepsilon)}\big(e_{rj}(1,\varepsilon) + e_{rr}(1,\varepsilon)\,\frac{p_{rj}(\varepsilon)}{1-p_{rr}(\varepsilon)}\big). \end{cases} \tag{3.39}$$

and

$$\begin{cases} {}_re_{rj}(2,\varepsilon) = \frac{1}{1-p_{rr}(\varepsilon)}\big(e_{rj}(2,\varepsilon) + e_{rr}(2,\varepsilon)\, {}_re_{rj}(0,\varepsilon) \\ \quad + 2e_{rr}(1,\varepsilon)\, {}_re_{rj}(1,\varepsilon)\big), \\ \quad = \frac{1}{1-p_{rr}(\varepsilon)}\big(e_{rj}(2,\varepsilon) + e_{rr}(2,\varepsilon)\frac{p_{rj}(\varepsilon)}{1-p_{rr}(\varepsilon)} \\ \quad + 2e_{rr}(1,\varepsilon)\frac{1}{1-p_{rr}(\varepsilon)}\big(e_{rj}(1,\varepsilon) + e_{rr}(1,\varepsilon)\frac{p_{rj}(\varepsilon)}{1-p_{rr}(\varepsilon)}\big)\big), \\ {}_re_{ij}(2,\varepsilon) = e_{ij}(2,\varepsilon) + e_{ir}(2,\varepsilon)\, {}_re_{rj}(0,\varepsilon) + 2e_{ir}(1,\varepsilon)\, {}_re_{rj}(1,\varepsilon) \\ \quad + \frac{p_{ir}(\varepsilon)}{1-p_{rr}(\varepsilon)}\big(e_{rj}(2,\varepsilon) + e_{rr}(2,\varepsilon)\, {}_re_{rj}(0,\varepsilon) \\ \quad + 2e_{rr}(1,\varepsilon)\, {}_re_{rj}(1,\varepsilon)\big) \\ \quad = e_{ij}(2,\varepsilon) + e_{ir}(2,\varepsilon)\frac{p_{rj}(\varepsilon)}{1-p_{rr}(\varepsilon)} \\ \quad + 2e_{ir}(1,\varepsilon)\frac{1}{1-p_{rr}(\varepsilon)}\big(e_{rj}(1,\varepsilon) + e_{rr}(1,\varepsilon)\frac{p_{rj}(\varepsilon)}{1-p_{rr}(\varepsilon)}\big), \\ \quad + \frac{p_{ir}(\varepsilon)}{1-p_{rr}(\varepsilon)}\big(e_{rj}(2,\varepsilon) + e_{rr}(2,\varepsilon)\frac{p_{rj}(\varepsilon)}{1-p_{rr}(\varepsilon)} \\ \quad + 2e_{rr}(1,\varepsilon)\frac{1}{1-p_{rr}(\varepsilon)}\big(e_{rj}(1,\varepsilon) + e_{rr}(1,\varepsilon)\frac{p_{rj}(\varepsilon)}{1-p_{rr}(\varepsilon)}\big)\big). \end{cases} \tag{3.40}$$

Again, it is useful to note that the second formula in relation (3.39) reduces to the first one, if to assign $i = r$ in this formula.

3.2.2 Basic Model Conditions for Reduced Semi-Markov Processes

Let us introduce the following sets, for $i, r \in \mathbb{X}$:

$$\mathbb{Y}_{ir}^{+} = \{j \in {}_r\mathbb{X} : j \in \mathbb{Y}_i\} \text{ and } \mathbb{Y}_{ir}^{-} = \begin{cases} \{j \in {}_r\mathbb{X} : j \in \mathbb{Y}_r\} & \text{if } r \in \mathbb{Y}_i, \\ \emptyset & \text{if } r \notin \mathbb{Y}_i. \end{cases} \tag{3.41}$$

Lemma 3.2. *Let the initial distribution $\bar{p}_\varepsilon$ be concentrated on the phase space ${}_r\mathbb{X}$, i.e., $p_{\varepsilon,r} = 0$. In this case, condition* **A**, *assumed to hold for the Markov chains $\eta_n^{(\varepsilon)}$, also holds for the Markov chains ${}_r\eta_n^{(\varepsilon)}$, with the same parameter ε_0 and transition sets ${}_r\mathbb{Y}_i$ defined by the following relation, for $i \in {}_r\mathbb{X}$:*

$${}_r\mathbb{Y}_i = \{j \in {}_r\mathbb{X} : {}_rp_{ij}(\varepsilon) > 0, \varepsilon \in (0, \varepsilon_0]\} = \mathbb{Y}_{ir}^{-} \cup \mathbb{Y}_{ir}^{+}. \tag{3.42}$$

Proof. Let $i \in {}_r\mathbb{X}$. If $j \in \mathbb{Y}_{ir}^{+}$, then $p_{ij}(\varepsilon) > 0$ and, thus, ${}_rp_{ij}(\varepsilon) > 0$. If $j \in \mathbb{Y}_{ir}^{-}$ then $p_{ir}(\varepsilon), p_{rj}(\varepsilon) > 0$ and, again, ${}_rp_{ij}(\varepsilon) > 0$. If $j \notin \mathbb{Y}_{ir}^{+} \cup \mathbb{Y}_{ir}^{-}$, then $p_{ij}(\varepsilon) = 0$ and $p_{ir}(\varepsilon) \cdot p_{rj}(\varepsilon) = 0$, and, thus, ${}_rp_{ij}(\varepsilon) = 0$. Therefore, relation (3.42) holds. If $\mathbb{Y}_{ir}^{+} \neq \emptyset$, then ${}_r\mathbb{Y}_i \neq \emptyset$. If $\mathbb{Y}_{ir}^{+} = \emptyset$, then $r \in \mathbb{Y}_i$ and, thus, $p_{ir}(\varepsilon) > 0$. Then, $\mathbb{Y}_{ir}^{-} = \{j \in {}_r\mathbb{X} : p_{rj}(\varepsilon) > 0\} = \mathbb{Y}_{rr}^{+} \neq \emptyset$. Therefore, sets ${}_r\mathbb{Y}_i \neq \emptyset, i \in {}_r\mathbb{X}$. Thus, conditions **A (a)** and **(b)** assumed to hold for the Markov chain $\eta_n^{(\varepsilon)}$, imply that these conditions also hold for the Markov chain ${}_r\eta_n^{(\varepsilon)}$, with sets ${}_r\mathbb{Y}_i, i \in {}_r\mathbb{X}$ replacing sets $\mathbb{Y}_i, i \in \mathbb{X}$.

Also, let $i, j \in {}_r\mathbb{X}$ and $i = l_0, l_1, \ldots, l_{n_{ij}} = j$ be a chain of states such that $l_1 \in \mathbb{Y}_{l_0}, \ldots, l_n \in \mathbb{Y}_{l_{n_{ij}-1}}$. We can always to assume that states $l_1, \ldots, l_{n_{ij}-1}$ are different and that $l_1, \ldots, l_{n_{ij}-1} \neq i, j$. This implies that either $l_1, \ldots, l_{n_{ij}-1} \neq r$ or there exist at most one $1 \leq k \leq n_{ij} - 1$ such that $i_k = r$. In the first case, $l_1 \in {}_r\mathbb{Y}_{l_0}, \ldots, l_{n_{ij}} \in {}_r\mathbb{Y}_{l_{n_{ij}-1}}$. In the second case, $l_1 \in {}_r\mathbb{Y}_{l_0}, \ldots, l_{k-1} \in {}_r\mathbb{Y}_{l_{k-2}}, l_{k-1} \in {}_r\mathbb{Y}_{l_{k+1}}, \ldots, l_{n_{ij}} \in {}_r\mathbb{Y}_{l_{n_{ij}-1}}$. Thus, condition **A (c)** assumed to hold for the Markov chain $\eta_n^{(\varepsilon)}$, imply that this condition also holds for the Markov chain ${}_r\eta_n^{(\varepsilon)}$. □

Remark 3.1. Relation (3.42) also holds for the case $i = r$. In this case sets $\mathbb{Y}_{rr}^- \subseteq \mathbb{Y}_{rr}^+$ and, thus, set ${}_r\mathbb{Y}_r = \mathbb{Y}_{ir}^- \cup \mathbb{Y}_{rr}^+ = \mathbb{Y}_{rr}^+ = \mathbb{Y}_r \setminus \{r\} \neq \emptyset$. Indeed, by condition **A**, probability $1 - p_{\varepsilon,rr} > 0$, for $r \in \mathbb{X}$, $\varepsilon \in (0, \varepsilon_0]$.

Lemma 3.3. *Conditions* **B** *and* **C***, assumed to hold for the semi-Markov processes* $\eta^{(\varepsilon)}(t)$*, also hold for the semi-Markov processes* ${}_r\eta_\varepsilon(t)$.

Proof. It directly follows from relations (3.30) and (3.38).

3.2.3 Hitting Times for Reduced Semi-Markov Processes

Let us define hitting times for the reduced semi-Markov process ${}_r\eta_\varepsilon(t)$, which are defined by the relation analogous to (3.5), for $j \in {}_r\mathbb{X}$,

$$ {}_r\tau_{\varepsilon,j} = \sum_{n=1}^{{}_r\nu_{\varepsilon,j}} {}_r\kappa_{\varepsilon,n}, \text{ where } {}_r\nu_{\varepsilon,j} = \min(n \geq 1 : {}_r\eta_{\varepsilon,n} = j). \tag{3.43} $$

The following theorem, analogous to those given in Silvestrov and Manca (2017), plays an important role in what follows.

Theorem 3.1. *Let conditions* **A**, **B**, *and* $\mathbf{C}_d$ *hold for semi-Markov processes* $\eta_\varepsilon(t)$. *Then, for any states* $r \in \mathbb{X}$ *and* $j \in {}_r\mathbb{X}$, *the first hitting times* $\tau_{\varepsilon,j}$ *and* ${}_r\tau_{\varepsilon,j}$ *to the state* j, *respectively, for semi-Markov processes* $\eta_\varepsilon(t)$ *and* ${}_r\eta_\varepsilon(t)$, *coincide, and, thus, for every* $k = 1, \ldots, d$, $j \in {}_r\mathbb{X}$, $i, r \in \mathbb{X}$ *and* $\varepsilon \in (0, \varepsilon_0]$, *the moments of hitting times* $E_{ij}(k, \varepsilon) = \mathsf{E}_i \tau_{\varepsilon,j}^k = \mathsf{E}_i \, {}_r\tau_{\varepsilon,j}^k$.

Proof. The first hitting times to a state $j \in {}_r\mathbb{X}$ are connected for Markov chains $\eta_{\varepsilon,n}$ and ${}_r\eta_{\varepsilon,n}$ by the following relation:

$$ \nu_{\varepsilon,j} = \min(n \geq 1 : \eta_{\varepsilon,n} = j) = \min({}_r\xi_{\varepsilon,n} \geq 1 : {}_r\eta_{\varepsilon,n} = j) = {}_r\xi_{\varepsilon,{}_r\nu_{\varepsilon,j}}, \tag{3.44} $$

where ${}_r\nu_{\varepsilon,j} = \min(n \geq 1 : {}_r\eta_{\varepsilon,n} = j)$.

The above relations imply that the following relation holds for the first hitting times to a state $j \in {}_r\mathbb{X}$, for the semi-Markov processes $\eta^{(\varepsilon)}(t)$ and ${}_r\eta_\varepsilon(t)$,

$$ \tau_{\varepsilon,j} = \sum_{n=1}^{\nu_{\varepsilon,j}} \kappa_{\varepsilon,n} = \sum_{n=1}^{{}_r\xi_{\varepsilon,{}_r\nu_{\varepsilon,j}}} \kappa_{\varepsilon,n} = \sum_{n=1}^{{}_r\nu_{\varepsilon,j}} {}_r\kappa_{\varepsilon,n} = {}_r\tau_{\varepsilon,j}. \tag{3.45} $$

The equality of power moments is an obvious corollary of relation (3.45). □

3.3 Asymptotic Expansions for Moments of Hitting Times for Perturbed Semi-Markov Processes

In this section, we present algorithms for computing of asymptotic expansions for transition characteristics of nonlinearly perturbed semi-Markov processes with reduced phase spaces and algorithms for sequential phase space reduction for semi-Markov processes and construction of Laurent asymptotic expansions, without and with explicit upper bound for remainders, for power moment of hitting times.

We would like to preface Lemmas 3.4–3.5 and Theorems 3.2–3.7, presenting these algorithms, by comments clarifying slightly unusual references in formulations of the above lemmas and theorems to descriptions of algorithms given in their proofs.

All lemmas and theorems mentioned above contain proofs of propositions that the corresponding functionals for perturbed reduced semi-Markov processes can be represented in the form of asymptotic expansions. These proofs are based on recurrent application of the operational rules for Laurent asymptotic expansions presented in Section 2.1 to the reduced semi-Markov processes constructed with the use of the corresponding recurrent time-space screening procedures of phase space reduction. In fact, one should correctly describe to which functions, in which order, and which operational rules should be applied for getting the corresponding expansions (their parameters, coefficients, and remainders) as well as to indicate some particular cases, where the corresponding computational steps should be modified. This is exactly what is done in the proofs of the corresponding lemmas and theorems.

An explicit writing down of the corresponding operational formulas representing the above recurrent algorithms (which could be given, say, as corollaries of these lemmas and theorems) would, in fact, replicate the above proofs in the formal form, require implementation of a huge number of intermediate notations, take too much space, etc., but would not add any new essential information about the corresponding algorithms. That is why the decision was made, just, to say in each lemma or theorem that the description of the corresponding algorithm is given in its proof. This makes formulations slightly unusual. But, as we think, this is the most compact way for presentation of the corresponding asymptotic results and algorithms.

3.3.1 Asymptotic Expansions with Remainders Given in the Standard Form for Transition Characteristics of Reduced Semi-Markov Processes

As was mentioned above, condition **A** implies that sets $\mathbb{Y}_{rr}^{+} \neq \emptyset, r \in \mathbb{X}$ and the non-absorption probability $\bar{p}_{rr}(\varepsilon) = 1 - p_{rr}(\varepsilon) > 0$, for $r \in \mathbb{X}$ and $\varepsilon \in (0, \varepsilon_0]$. This probability satisfies the following relation, for every $r \in \mathbb{X}$ and $\varepsilon \in (0, \varepsilon_0]$:

$$\bar{p}_{rr}(\varepsilon) = 1 - p_{rr}(\varepsilon) = \sum_{j \in \mathbb{Y}_{rr}^+} p_{rj}(\varepsilon). \tag{3.46}$$

Lemma 3.4. *Let conditions **A** and **D** hold. Then, the pivotal $(\bar{l}_{rr}^-, \bar{l}_{rr}^+)$-expansions for the non-absorption probabilities $\bar{p}_{rr}(\varepsilon), r \in \mathbb{X}$ are given by the algorithm described below, in the proof of the lemma.*

Proof. Let $r \in \mathbb{Y}_r$. First, proposition **(i)** (the multiple summation rule) of Lemma 2.3 should be applied to the sum $\sum_{j \in \mathbb{Y}_{rr}^+} p_{rj}(\varepsilon)$.

Second, propositions **(i)** (the multiplication by constant -1) and **(ii)** (the summation with constant 1) of Lemma 2.2 should be applied to the asymptotic expansion for probability $p_{rr}(\varepsilon)$ given in condition **B**, in order to get the asymptotic expansion for function $1 - p_{rr}(\varepsilon)$.

Third, Lemma 1 should be applied to the asymptotic expansion for function $\bar{p}_{rr}(\varepsilon)$ given in two alternative forms by relation (3.46).

Note that condition **F** holds also for the above case, where the asymptotic expansion for probability $\bar{p}_{rr}(\varepsilon)$, obtained at the second step, is replaced by the improved version of this expansion, obtained with the use of Lemma 2.1 at the third step.

The case $r \notin \mathbb{Y}_r$ is trivial, since, in this case, probability $\bar{p}_{rr}(\varepsilon) \equiv 1$.

According to Lemmas 2.1–2.3, $(\bar{l}_{rr}^-, \bar{l}_{rr}^+)$-expansions,

$$\bar{p}_{rr}(\varepsilon) = \sum_{l=\bar{l}_{rr}^-}^{\bar{l}_{rr}^+} \bar{a}_{rr}[l]\varepsilon^l + \bar{o}_{rr}(\varepsilon^{\bar{l}_{rr}^+}), \varepsilon \in (0, \varepsilon_0], \tag{3.47}$$

yielded, for every $r \in \mathbb{X}$, by the above algorithm, are pivotal. □

Theorem 3.2. *Let the initial distribution $\bar{p}_\varepsilon$ is concentrated on the phase space ${}_r\mathbb{X}$, i.e., $p_{\varepsilon,r} = 0$. In this case, conditions **A** and **D**, assumed to hold for the Markov chains $\eta_{\varepsilon,n}$, also hold for the reduced Markov chains ${}_r\eta_{\varepsilon,n}$, with the same parameter ε_0 and the transition sets ${}_r\mathbb{Y}_i, i \in {}_r\mathbb{X}$, given by relation (3.42). The pivotal $({}_rl_{ij}^-, {}_rl_{ij}^+)$-expansions appearing in condition **D** are given for transition probabilities ${}_rp_{ij}(\varepsilon), j \in {}_r\mathbb{Y}_i, i \in {}_r\mathbb{X}, r \in \mathbb{X}$ by the algorithm described below, in the proof of the theorem.*

Proof. Lemma 3.2 implies that condition **A** holds for the Markov chains ${}_r\eta_{\varepsilon,n}$, with the same parameter ε_0 as for the Markov chains $\eta_{\varepsilon,n}$, and the transition sets ${}_r\mathbb{Y}_i, i \in {}_r\mathbb{X}$ given by relation (3.42).

Let us prove that condition **D** also holds for the Markov chains ${}_r\eta_{\varepsilon,n}$, with the same parameter ε_0 and the transition sets ${}_r\mathbb{Y}_i, i \in {}_r\mathbb{X}$ given by relation (3.42). In order to do this, let us construct the corresponding asymptotic expansions appearing in this condition.

Let $j, r \in \mathbb{Y}_i \cap \mathbb{Y}_r$. We construct the asymptotic expansions for probabilities ${}_rp_{ij}(\varepsilon)$ using formulas (3.31) and the asymptotic expansions appearing in condition **D**.

First, proposition **(v)** (the division rule) of Lemma 2.2 should be applied to the quotient ${}_rp_{rj}(\varepsilon) = \frac{p_{rj}(\varepsilon)}{1 - p_{rr}(\varepsilon)}$.

Second, proposition **(iii)** (the multiplication rule) of Lemma 2.2 should be applied to the product $p_{ir}(\varepsilon) \cdot {}_r p_{rj}(\varepsilon)$.

Third, proposition **(ii)** (the summation rule) of Lemma 2.2 should be applied to sum ${}_r p_{ij}(\varepsilon) = p_{ij}(\varepsilon) + p_{ir}(\varepsilon) \cdot {}_r p_{ij}(\varepsilon)$.

The asymptotic expansions for probabilities $p_{ij}(\varepsilon)$, $p_{ir}(\varepsilon)$, and $p_{rj}(\varepsilon)$, given in condition **D** are used in the algorithm described above.

If $j \notin \mathbb{Y}_i$, then $p_{ij}(\varepsilon) \equiv 0$; if $j \notin \mathbb{Y}_r$, then $p_{rj}(\varepsilon) \equiv 0$; if $r \notin \mathbb{Y}_i$, then $p_{ir}(\varepsilon) \equiv 0$; if $r \notin \mathbb{Y}_r$, then $1 - p_{rr}(\varepsilon) \equiv 1$. In these cases, the above algorithm is readily simplified, with the use of Lemma 2.4.

Note that parameter ε_0 does not change in the multiplication and summation steps as well as in the division step, since $1 - p_{rr}(\varepsilon) > 0$, $\varepsilon \in (0, \varepsilon_0]$.

According to Lemma 2.2, the $({}_r l^-_{ij}, {}_r l^+_{ij})$-expansions,

$$
{}_r p_{ij}(\varepsilon) = \sum_{l={}_r l^-_{ij}}^{{}_r l^+_{ij}} {}_r a_{ij}[l]\varepsilon^l + {}_r o_{ij}(\varepsilon^{{}_r l^+_{ij}}), \varepsilon \in (0, \varepsilon_0], \tag{3.48}
$$

yielded, for $j \in {}_r\mathbb{Y}_i, i \in {}_r\mathbb{X}, r \in \mathbb{X}$, by the above algorithm, are pivotal. □

Remark 3.2. The matrix of transition probabilities $\|{}_r p_{ij}(\varepsilon)\|$ is stochastic, for every $\varepsilon \in (0, \varepsilon_0]$. Thus, under conditions of Theorem 3.2, condition **F** holds for the asymptotic expansions of transition probabilities ${}_r p_{ij}(\varepsilon)$, $j \in {}_r Y_i, i \in {}_r\mathbb{X}$ given in this theorem, for every $r \in \mathbb{X}$.

Remark 3.3. The algorithm described in the proof of Theorem 3.2 also yields the pivotal asymptotic expansions for the transition probabilities ${}_r p_{rj}(\varepsilon), j \in {}_r\mathbb{Y}_r, r \in \mathbb{X}$. They, actually, are obtained at the first step the algorithm presented in the above proof and take the form given in relation (3.48), where one should assign $i = r$.

Theorem 3.3. *Let the initial distribution $\bar{p}_\varepsilon$ is concentrated on the phase space ${}_r\mathbb{X}$, i.e., $p_{\varepsilon,r} = 0$. In this case, conditions* **A**, **B**, $\mathbf{C}_d$, **D**, *and* $\mathbf{E}_d$, *assumed to hold for the semi-Markov processes $\eta_\varepsilon(t)$, also hold for the reduced semi-Markov processes ${}_r\eta_\varepsilon(t)$. Parameter ε_0, in conditions* **A**, **D**, *and* $\mathbf{E}_d$, *is the same for processes $\eta_\varepsilon(t)$ and ${}_r\eta_\varepsilon(t)$. The transition sets ${}_r\mathbb{Y}_i, i \in {}_r\mathbb{X}$ are given for processes ${}_r\eta_\varepsilon(t)$ by relation* (3.42). *The pivotal $({}_r m^-_{ij}[k], {}_r m^+_{ij}[k])$-expansions appearing in condition* $\mathbf{E}_d$ *are given for expectations ${}_r e_{ij}(k, \varepsilon), k = 1, \ldots, d, j \in {}_r\mathbb{Y}_i, i \in {}_r\mathbb{X}, r \in \mathbb{X}$ by the recurrent algorithm described below, in the proof of the theorem.*

Proof. Lemma 3.2 and Theorem 3.2 imply that conditions **A** and **D** hold for the semi-Markov processes ${}_r\eta_\varepsilon(t)$, with the same parameter ε_0 as for the semi-Markov processes $\eta_\varepsilon(t)$, and the transition sets ${}_r\mathbb{Y}_i, i \in {}_r\mathbb{X}$ given by relation (3.42). Also, conditions **B** and $\mathbf{C}_d$ hold for the semi-Markov processes ${}_r\eta_\varepsilon(t)$, by Lemma 3.3.

In order to prove that condition $\mathbf{E}_d$ also holds for the semi-Markov processes ${}_r\eta_\varepsilon(t)$, with the same parameter ε_0 and the transition sets ${}_r\mathbb{Y}_i, i \in {}_r\mathbb{X}$ given by relation (3.42), let us construct the corresponding asymptotic expansions appearing in this condition.

Let $j, r \in \mathbb{Y}_i \cap \mathbb{Y}_r$. At the first recurrent step, we construct the asymptotic expansions for expectations ${}_re_{rj}(1,\varepsilon)$ and ${}_re_{ij}(1,\varepsilon)$ using formulas (3.39) and the corresponding asymptotic expansions appearing in conditions **D** and $\mathbf{E}_d$.

First, proposition **(iii)** (the multiplication rule) of Lemma 2.2 should be applied to the product $e_{rr}(1,\varepsilon) \cdot {}_re_{rj}(0,\varepsilon)$. Here, the asymptotic expansion for probability ${}_re_{rj}(0,\varepsilon) = {}_rp_{rj}(\varepsilon)$ constructed in Lemma 3.2 should be used.

Second, proposition **(ii)** (the summation rule) of Lemma 2.2 should be applied to the sum $e_{rj}(1,\varepsilon) + e_{rr}(1,\varepsilon) \cdot {}_re_{rj}(0,\varepsilon)$.

Third, proposition **(v)** (the division rule) of Lemma 2.2 should be applied to the quotient ${}_re_{rj}(1,\varepsilon) = \frac{1}{1-p_{rr}(\varepsilon)}(e_{rj}(1,\varepsilon) + e_{rr}(1,\varepsilon) \cdot {}_re_{rj}(0,\varepsilon))$. Here, the asymptotic expansion for probability $1 - p_{rr}(\varepsilon)$ constructed in Lemma 3.4 should be used.

Fourth, proposition **(iii)** (the multiplication rule) of Lemma 2.2 should be applied to the products $e_{ir}(1,\varepsilon) \cdot {}_re_{rj}(0,\varepsilon)$, and $p_{ir}(\varepsilon) \cdot {}_re_{rj}(1,\varepsilon)$. Here, the asymptotic expansion for expectations ${}_re_{rj}(0,\varepsilon) = {}_rp_{rj}(\varepsilon)$ constructed in Theorem 3.2 and ${}_re_{rj}(1,\varepsilon)$ constructed above should be used.

Fifth, the proposition **(i)** (the multiple summation rule) of Lemma 2.3 should be applied to sum ${}_re_{ij}(1,\varepsilon) = e_{ij}(1,\varepsilon) + e_{ir}(1,\varepsilon) \cdot {}_re_{rj}(0,\varepsilon) + p_{ir}(\varepsilon) \cdot {}_re_{rj}(1,\varepsilon)$.

In this case, the asymptotic expansions for probabilities $p_{ij}(\varepsilon), p_{ir}(\varepsilon)$, and $p_{rj}(\varepsilon)$, given in condition **D**, and expectations $e_{ij}(1,\varepsilon), e_{ir}(1,\varepsilon), e_{rr}(1,\varepsilon)$, and $e_{rj}(1,\varepsilon)$, given in condition $\mathbf{E}_d$, are used in the first recurrent step of the algorithm described above.

If $j \notin \mathbb{Y}_i$, then $p_{ij}(\varepsilon) \equiv 0$ and $e_{ij}(1,\varepsilon) \equiv 0$; if $j \notin \mathbb{Y}_r$, then $p_{rj}(\varepsilon) \equiv 0$ and $e_{rj}(1,\varepsilon) \equiv 0$; if $r \notin \mathbb{Y}_i$, then $p_{ir}(\varepsilon) \equiv 0$ and $e_{ir}(1,\varepsilon) \equiv 0$; if $r \notin \mathbb{Y}_r$, then $1 - p_{rr}(\varepsilon) \equiv 1$ and $e_{rr}(1,\varepsilon) \equiv 0$. In these cases, the above algorithm is readily simplified, with the use of Lemma 2.4.

Note that parameter ε_0 does not change in the multiplication and summation steps as well as in the division step, since $1 - p_{rr}(\varepsilon) > 0, \varepsilon \in (0, \varepsilon_0]$.

According to Lemma 2.2, the $({}_rm^-_{ij}[1], {}_rm^+_{ij}[1])$-expansions,

$$
{}_re_{ij}(1,\varepsilon) = \sum_{l={}_rm^-_{ij}[1]}^{{}_rm^+_{ij}[1]} {}_rb_{ij}[1,l]\varepsilon^l + {}_r\dot{o}_{1,ij}(\varepsilon^{{}_rm^+_{ij}[1]}), \varepsilon \in (0,\varepsilon_0], \tag{3.49}
$$

yielded, for $j \in {}_r\mathbb{Y}_i, i \in {}_r\mathbb{X}, r \in \mathbb{X}$, by the above algorithm, are pivotal.

The algorithm described above also yields the pivotal asymptotic expansions for expectations ${}_re_{rj}(1,\varepsilon), j \in {}_r\mathbb{Y}_r, r \in \mathbb{X}$. They, actually, are obtained at the third step of the above algorithm and take the form given (3.49), where one should assign $i = r$.

Five steps of the algorithm described above should be recurrently repeated for $k = 2, \ldots, d$.

Let assume that the corresponding pivotal asymptotic expansions for moments ${}_re_{rj}(l,\varepsilon)$, ${}_re_{ij}(l,\varepsilon), l = 1, \ldots, k-1$ have been already constructed with the use of formulas (3.38) and the corresponding asymptotic expansions appearing in conditions **D** and $\mathbf{E}_d$. In this case, the asymptotic expansions for moments ${}_re_{rj}(k,\varepsilon)$, ${}_re_{ij}(k,\varepsilon)$ can be constructed using the above asymptotic expansions, formulas (3.38), and the corresponding asymptotic expansions appearing in conditions **D** and $\mathbf{E}_d$, in the following way.

First, propositions **(i)** (the multiplication by constant rule) and **(iii)** (the multiplication rule) of Lemma 2.2 should be applied to the products $\binom{k}{l} e_{ir}(k-l,\varepsilon) \cdot {}_r e_{rj}(l,\varepsilon), l = 0,\dots,k-1$.

Second, proposition **(i)** (the multiple summation rule) of Lemma 2.3 should be applied to the sum $e_{rj}(k,\varepsilon) + \sum_{l=0}^{k-1} \binom{k}{l} e_{rr}(k-l,\varepsilon) \cdot {}_r e_{rj}(l,\varepsilon)$

Third, proposition **(v)** (the division rule) of Lemma 2.2 should be applied to the quotient ${}_r e_{rj}(k,\varepsilon) = \frac{1}{1-p_{rr}(\varepsilon)} (e_{rj}(k,\varepsilon) + \sum_{l=0}^{k-1} \binom{k}{l} e_{rr}(k-l,\varepsilon) \cdot {}_r e_{rj}(l,\varepsilon))$.

Fourth, propositions **(i)** (the multiplication by constant rule) and **(iii)** (the multiplication rule) of Lemma 2.2 should be applied to the products $\binom{k}{l} e_{ir}(k-l,\varepsilon) \cdot {}_r e_{rj}(l,\varepsilon), l = 0, k-1$ and $p_{ir}(\varepsilon) \cdot {}_r e_{rj}(k,\varepsilon)$. Here, the asymptotic expansion for moment ${}_r e_{rj}(k,\varepsilon)$ constructed above should be used.

Fifth, the proposition **(i)** (the multiple summation rule) of Lemma 2.3 should be applied to sum ${}_r e_{ij}(k,\varepsilon) = e_{ij}(k,\varepsilon) + \sum_{l=0}^{k-1} \binom{k}{l} e_{ir}(k-l,\varepsilon) \cdot {}_r e_{rj}(l,\varepsilon) + p_{ir}(\varepsilon) \cdot {}_r e_{rj}(k,\varepsilon)$.

In this case, the asymptotic expansions for probabilities $p_{ij}(\varepsilon), p_{ir}(\varepsilon)$, and $p_{rj}(\varepsilon)$, given in condition **D** and expectations $e_{ij}(l,\varepsilon), e_{ir}(l,\varepsilon), e_{rr}(l,\varepsilon), e_{rj}(l,\varepsilon), l = 1,\dots,k$, given in condition $\mathbf{E}_d$, are used at the k-th recurrent step of the algorithm.

As it was already mentioned above, five steps of the above algorithm should be recurrently repeated for $k = 1,2,\dots,d$.

If $j \notin \mathbb{Y}_i$, then $p_{ij}(\varepsilon) \equiv 0$ and $e_{ij}(k,\varepsilon) \equiv 0, k = 1,\dots,d$; if $j \notin \mathbb{Y}_r$, then $p_{rj}(\varepsilon) \equiv 0$ and $e_{rj}(k,\varepsilon) \equiv 0, k = 1,\dots,d$; if $r \notin \mathbb{Y}_i$, then $p_{ir}(\varepsilon) \equiv 0$ and $e_{ir}(k,\varepsilon) \equiv 0, k = 1,\dots,d$; if $r \notin \mathbb{Y}_r$, then $1 - p_{rr}(\varepsilon) \equiv 1$ and $e_{rr}(k,\varepsilon) \equiv 0, k = 1,\dots,d$. In these cases, the above recurrent algorithm is readily simplified with the use of Lemma 2.4.

As in Theorem 3.2, parameter ε_0 does not change in the multiplication and summation steps as well as in the division step, since $1 - p_{rr}(\varepsilon) > 0, \varepsilon \in (0,\varepsilon_0]$.

According to Lemmas 2.2 and 2.3, the $({}_r m_{ij}^-[k], {}_r m_{ij}^+[k])$-expansions,

$$ {}_r e_{ij}(k,\varepsilon) = \sum_{l={}_r m_{ij}^-[k]}^{{}_r m_{ij}^+[k]} {}_r b_{ij}[k,l]\varepsilon^l + {}_r\dot{o}_{k,ij}(\varepsilon^{{}_r m_{ij}^+[k]}), \varepsilon \in (0,\varepsilon_0], \tag{3.50} $$

yielded, for $k = 1,\dots,d, j \in {}_r\mathbb{Y}_i, i \in {}_r\mathbb{X}, r \in \mathbb{X}$, by the above algorithm, are pivotal. □

Remark 3.4. The algorithm described in the proof of Theorem 3.3 also yields the pivotal asymptotic expansions for moments ${}_r e_{rj}(k,\varepsilon), k = 1,\dots,d, j \in {}_r\mathbb{Y}_r, r \in \mathbb{X}$, which take the form given in relation (3.50), where one should assign $i = r$.

It is worth to note that, despite bulky forms, recurrent formulas for parameters and algorithms for computing coefficients in the asymptotic expansions, presented in Lemma 3.4 and Theorems 3.2 and 3.3, are computationally effective.

3.3.2 Asymptotic Expansions with Remainders Given in the Standard Form for Moments of Hitting Times for Reduced Semi-Markov Processes

In what follows, $i, j \in \mathbb{X}, i \neq j$ and let $\bar{r}_{i,j,N} = \langle r_{i,j,1}, \ldots, r_{i,j,N} \rangle = \langle r_{i,1}, \ldots, r_{i,N-2}, i, j \rangle$ be a permutation of the sequence $\langle 1, \ldots, N \rangle$ such that $r_{i,j,N-1} = i, r_{i,j,N} = j$, and let $\bar{r}_{i,j,n} = \langle r_{i,j,1}, \ldots, r_{i,j,n} \rangle$, $n = 1, \ldots, N$ be the corresponding chain of growing sequences of states from space $\mathbb{X}$.

Theorem 3.4. *Let conditions* **A**, **B**, $\mathbf{C}_d$, **D**, *and* $\mathbf{E}_d$ *hold for semi-Markov processes* $\eta_\varepsilon(t)$. *Then, for every* $i, j \in \mathbb{X}, i \neq j$, *the pivotal* $(M^-_{i'j}[k], M^+_{i'j}[k])$-*expansions, for moments of hitting time* $E_{i'j}(k, \varepsilon), k = 1, \ldots, d, i' = i, j$ *are given by the algorithm based on the sequential exclusion of states* $r_{i,j,1}, \ldots, r_{i,j,N-2}, i$ *from the phase space* $\mathbb{X}$ *of the processes* $\eta_\varepsilon(t)$. *This algorithm is described below, in the proof of the theorem. The above* $(M^-_{i'j}[k], M^+_{i'j}[k])$-*expansions are invariant with respect to any permutation* $\bar{r}_{i,j,N} = \langle r_{i,j,1}, \ldots, r_{i,j,N-2}, i, j \rangle$ *of sequence* $\langle 1, \ldots, N \rangle$.

Proof. Let us assume that $p_{\varepsilon,i} + p_{\varepsilon,j} = 1$.

Denote as ${}_{\bar{r}_{i,j,0}}\eta_\varepsilon(t) = \eta_\varepsilon(t)$, the initial semi-Markov process. Let us exclude state $r_{i,j,1}$ from the phase space of semi-Markov process ${}_{\bar{r}_{i,j,0}}\eta^{(\varepsilon)}(t)$ using the time-space screening procedure described in Section 3.2. Let ${}_{\bar{r}_{i,j,1}}\eta_\varepsilon(t)$ be the corresponding reduced semi-Markov process. Note that the initial distribution of the process ${}_{\bar{r}_{i,j,1}}\eta_\varepsilon(t)$ is concentrated in states $i, j \neq r_1$. According to the remarks made in Subsection 3.2.1, we can consider ${}_{\bar{r}_{i,j,1}}\eta_\varepsilon(t)$ as a standard semi-Markov process with the phase space ${}_{\bar{r}_{i,j,1}}\mathbb{X} = \mathbb{X} \setminus \{r_{i,j,1}\}$.

The above procedure can be repeated. The state $r_{i,j,2}$ can be excluded from the phase space of the semi-Markov process ${}_{\bar{r}_{i,j,1}}\eta^{(\varepsilon)}(t)$. Let ${}_{\bar{r}_{i,j,2}}\eta^{(\varepsilon)}(t)$ be the corresponding reduced semi-Markov process. By continuing the above procedure for states $r_{i,j,3}, \ldots, r_{i,j,n}$, we construct the reduced semi-Markov process ${}_{\bar{r}_{i,j,n}}\eta_\varepsilon(t)$.

We can consider ${}_{\bar{r}_{i,n}}\eta_\varepsilon(t)$ as a standard semi-Markov process with the phase space ${}_{\bar{r}_{i,j,n}}\mathbb{X} = \mathbb{X} \setminus \{r_{i,j,1}, r_{i,j,2}, \ldots, r_{i,j,n}\}$.

The transition probabilities of the embedded Markov chain ${}_{\bar{r}_{i,j,n}}p_{i'j'}(\varepsilon)$, $i', j' \in {}_{\bar{r}_{i,j,n}}\mathbb{X}$, and the moments of transition times ${}_{\bar{r}_{i,j,n}}e_{i'j'}(k, \varepsilon), k = 1, \ldots, d, i', j' \in {}_{\bar{r}_{i,n}}\mathbb{X}$ are determined for the semi-Markov process ${}_{\bar{r}_{i,j,n}}\eta_\varepsilon(t)$ by the transition probabilities and the corresponding moments of transition times for the process ${}_{\bar{r}_{i,j,n-1}}\eta_\varepsilon(t)$, respectively, via relations (3.31) and (3.38). Relations (3.31) and (3.38) also give formulas for probabilities ${}_{\bar{r}_{i,j,n}}p_{r_{i,j,n-1}j'}(\varepsilon)$, $j' \in {}_{\bar{r}_{i,j,n}}\mathbb{X}$ and moments ${}_{\bar{r}_{i,j,n}}e_{r_{i,j,n-1}j'}(k, \varepsilon), k = 1, \ldots, d$, $j' \in {}_{\bar{r}_{i,j,n}}\mathbb{X}$ determined by the transition probabilities and the corresponding moments of transition times for the process ${}_{\bar{r}_{i,j,n-1}}\eta_\varepsilon(t)$.

By Theorem 3.1, the moment of hitting time $E_{i'j'}(k, \varepsilon)$ coincides for the semi-Markov processes ${}_{\bar{r}_{i,j,0}}\eta_\varepsilon(t)$, ${}_{\bar{r}_{i,j,1}}\eta_\varepsilon(t), \ldots, {}_{\bar{r}_{i,j,n}}\eta^{(\varepsilon)}(t)$, for every $k = 1, \ldots, d, i', j' \in {}_{\bar{r}_{i,j,n}}\mathbb{X}$.

By Theorems 3.2 and 3.3, the semi-Markov processes ${}_{\bar{r}_{i,j,n}}\eta_\varepsilon(t)$ satisfy conditions **B**, $\mathbf{C}_d$ and, also, conditions **A**, **D**, and $\mathbf{E}_d$, with the same parameter ε_0 as for processes ${}_{\bar{r}_{i,j,n-1}}\eta_\varepsilon(t)$. The transition sets ${}_{\bar{r}_{i,j,n}}\mathbb{Y}_{i'}, i' \in {}_{\bar{r}_{i,j,n}}\mathbb{X}$ determined by the transition sets ${}_{\bar{r}_{i,j,n-1}}\mathbb{Y}_{i'}, i' \in {}_{\bar{r}_{i,j,n-1}}\mathbb{X}$, via relation (3.42) given in Lemma 3.2.

Therefore, the pivotal $({}_{\bar{r}_{i,j,n}}l^-_{i'j'}, {}_{\bar{r}_{i,j,n}}l^+_{i'j'})$-expansions,

$$
{}_{\bar{r}_{i,j,n}}p_{i'j'}(\varepsilon) = \sum_{{}_{\bar{r}_{i,j,n}}l^-_{i'j'}}^{{}_{\bar{r}_{i,j,n}}l^+_{i'j'}} {}_{\bar{r}_{i,j,n}}a_{i'j'}[l]\varepsilon^l + {}_{\bar{r}_{i,j,n}}o_{i'j'}(\varepsilon^{{}_{\bar{r}_{i,j,n}}l^+_{i'j'}}), \tag{3.51}
$$

can be constructed for $j' \in {}_{\bar{r}_{i,j,n}}\mathbb{Y}_{i'}$, $i' \in {}_{\bar{r}_{i,j,n}}\mathbb{X}$ by applying the algorithms given in Theorem 3.2 to the $({}_{\bar{r}_{i,j,n-1}}l^-_{i'j'}, {}_{\bar{r}_{i,j,n-1}}l^+_{i'j'})$-expansions for transition probabilities ${}_{\bar{r}_{i,j,n-1}}p_{i'j'}(\varepsilon)$, $j' \in {}_{\bar{r}_{i,j,n-1}}\mathbb{Y}_{i'}$, $i' \in {}_{\bar{r}_{i,j,n-1}}\mathbb{X}$.

Also, the pivotal $({}_{\bar{r}_{i,j,n}}m^-_{i'j'}[k], {}_{\bar{r}_{i,j,n}}m^+_{i'j'}[k])$-expansions,

$$
{}_{\bar{r}_{i,j,n}}e_{i'j'}(k,\varepsilon) = \sum_{{}_{\bar{r}_{i,j,n}}m^-_{i'j'}[k]}^{{}_{\bar{r}_{i,j,n}}m^+_{i'j'}[k]} {}_{\bar{r}_{i,j,n}}b_{i'j'}[k,l]\varepsilon^l + {}_{\bar{r}_{i,j,n}}\dot{o}_{k,i'j'}(\varepsilon^{{}_{\bar{r}_{i,j,n}}m^+_{i'j'}[k]}), \tag{3.52}
$$

can be constructed for $k = 1,\dots,d$, $j' \in {}_{\bar{r}_{i,j,n}}\mathbb{Y}_{i'}, i' \in {}_{\bar{r}_{i,j,n}}\mathbb{X}$ by applying the algorithms given in Theorems 3.2 and 3.3 to the $({}_{\bar{r}_{i,j,n-1}}m^-_{i'j'}[k], {}_{\bar{r}_{i,j,n-1}}m^+_{i'j'}[k])$-expansions for moments ${}_{\bar{r}_{i,j,n-1}}e_{i'j'}(k,\varepsilon), k = 1,\dots,d, j' \in {}_{\bar{r}_{i,j,n-1}}\mathbb{Y}_{i'}, i' \in {}_{\bar{r}_{i,j,n-1}}\mathbb{X}$.

The algorithm described above has a recurrent form and should be realized sequentially for the reduced semi-Markov processes ${}_{\bar{r}_{i,j,1}}\eta_\varepsilon(t), \dots, {}_{\bar{r}_{i,j,n}}\eta_\varepsilon(t)$ starting from the initial semi-Markov process ${}_{\bar{r}_{i,j,0}}\eta_\varepsilon(t)$.

For every $j' \in {}_{\bar{r}_{i,j,n}}\mathbb{Y}_{i'}, i' \in {}_{\bar{r}_{i,j,n}}\mathbb{X}$, $n = 1,\dots,N-2$, the asymptotic expansions for the transition probability ${}_{\bar{r}_{i,j,n}}p_{i'j'}(\varepsilon)$ and the moments ${}_{\bar{r}_{i,j,n}}e_{i'j'}(k,\varepsilon)$, $k = 1,\dots,d$, resulted by the recurrent algorithm of sequential phase space reduction described above, are invariant with respect to any permutation $\bar{r}'_{i,j,n} = \langle r'_{i,j,1},\dots, r'_{i,j,n}\rangle$ of sequence $\bar{r}_{i,j,n} = \langle r_{i,j,1},\dots,r_{i,j,n}\rangle$.

Indeed, for every permutation $\bar{r}'_{i,j,n}$ of sequence $\bar{r}_{i,j,n}$, the corresponding reduced semi-Markov process ${}_{\bar{r}'_{i,j,n}}\eta_\varepsilon(t)$ is constructed as the sequence of states for the initial semi-Markov process $\eta_\varepsilon(t)$ at sequential moments of its hitting into the same reduced phase space ${}_{\bar{r}'_{i,j,n}}\mathbb{X} = \mathbb{X}\setminus\{r'_{i,j,1},\dots,r'_{i,j,n}\} = {}_{\bar{r}_{i,j,n}}\mathbb{X} = \mathbb{X}\setminus\{r_{i,j,1},\dots,r_{i,j,n}\}$. The times between sequential jumps of the reduced semi-Markov process ${}_{\bar{r}'_{i,j,n}}\eta_\varepsilon(t)$ are the times between sequential instants of hitting the above reduced phase space by the initial semi-Markov process $\eta_\varepsilon(t)$.

This implies that the transition probability ${}_{\bar{r}_{i,n}}p_{i'j'}(\varepsilon)$ and the moment ${}_{\bar{r}_{i,j,n}}e_{i'j'}(k,\varepsilon)$ are, for every $k = 1,\dots,d, j' \in {}_{\bar{r}_{i,n}}\mathbb{Y}_{i'}, i' \in {}_{\bar{r}_{i,n}}\mathbb{X}, n = 1,\dots,N-2$, invariant (as functions of ε) with respect to any permutation $\bar{r}'_{i,j,n}$ of the sequence $\bar{r}_{i,j,n}$. Moreover, as follows from algorithms presented above, in Lemma 2.4 and Theorems 3.2 and 3.3, the transition probability ${}_{\bar{r}_{i,j,n}}p_{i'j'}(\varepsilon)$ is a rational function of the initial transition probabilities $p_{ij}(\varepsilon), j \in \mathbb{Y}_i, i \in \mathbb{X}$, and the moment ${}_{\bar{r}_{i,j,n}}e_{i'j'}(k,\varepsilon)$ is a rational function of the initial transition probabilities $p_{i''j''}(\varepsilon), j'' \in \mathbb{Y}_{i''}, i'' \in \mathbb{X}$ and the moments of sojourn times $e_{i''j''}(l,\varepsilon), l = 1,\dots,k, j'' \in \mathbb{Y}_{i''}, i'' \in \mathbb{X}$ (quotients of sums of products for some of these probabilities and expectations). According to the above remarks, the transition probability ${}_{\bar{r}_{i,n}}p_{i'j'}(\varepsilon)$ and moment ${}_{\bar{r}_{i,j,n}}e_{i'j'}(k,\varepsilon)$ are invariant with respect to any permutation $\bar{r}'_{i,j,n}$ of the sequence $\bar{r}_{i,j,n}$.

By using identity arithmetical transformations (disclosure of brackets, imposition of a common factor out of the brackets, bringing a fractional expression to a common denominator, permutation of summands or multipliers, elimination of expressions with equal absolute values and opposite signs in the sums, and elimination of equal expressions in quotients) the rational functions ${}_{\bar{r}'_{i,j,n}}p_{i'j'}(\varepsilon)$ and ${}_{\bar{r}'_{i,j,n}}e_{i'j'}(k,\varepsilon)$ can be transformed, respectively, into the rational functions ${}_{\bar{r}_{i,j,n}}p_{i'j'}(\varepsilon)$ and ${}_{\bar{r}_{i,j,n}}e_{i'j'}(\varepsilon)$ and vice versa. By Lemma 2.4, these transformations do not affect the corresponding asymptotic expansions for functions ${}_{\bar{r}_{i,j,n}}p_{i'j'}(\varepsilon)$ and ${}_{\bar{r}_{i,j,n}}e_{i'j'}(k,\varepsilon)$ and, thus, these expansions are invariant with respect to any permutation $\bar{r}'_{i,j,n}$ of the sequence $\bar{r}_{i,j,n}$.

In fact, one should only check the above invariance propositions for the case, where the permutation $\bar{r}'_{i,j,n}$ is obtained from the sequence $\bar{r}_{i,j,n}$ by exchange of a pair of neighbor states $r_{i,j,l}$ and $r_{i,j,l+1}$, for some $1 \leq l \leq n-1$. Then, the proof can be repeated for a pair of neighbor states for the sequence $\bar{r}'_{i,j,n}$, etc. In this way, the proof can be expanded to the case of an arbitrary permutation $\bar{r}'_{i,j,n}$ of the sequence $\bar{r}_{i,j,n}$. The above-mentioned proof of pairwise permutation invariance involves processes ${}_{\bar{r}_{i,j,l-1}}\eta_\varepsilon(t)$, ${}_{\bar{r}_{i,j,l}}\eta_\varepsilon(t)$, and ${}_{\bar{r}_{i,j,l+1}}\eta_\varepsilon(t)$. It is absolutely analogous, for $1 \leq l \leq n-1$. Taking this into account, we just show how this proof can be accomplished for the case $l = 1$.

The transition probabilities ${}_{\bar{r}_{i,j,2}}p_{i'j'}(\varepsilon)$ and ${}_{\bar{r}'_{i,j,2}}p_{i'j'}(\varepsilon)$ for the sequences $\bar{r}_{i,j,2} = \langle r_1, r_2\rangle$ and $\bar{r}'_{i,j,2} = \langle r_2, r_1\rangle$ (here, $i, j, i', j' \neq r_1, r_2$) can be transformed into the same symmetric (with respect to r_1, r_2) rational function of the corresponding transition probabilities, using the identity arithmetical transformations listed above,

$$
\begin{aligned}
{}_{\bar{r}_{i,j,2}}p_{i'j'}(\varepsilon) &= {}_{r_1}p_{i'j'}(\varepsilon) + {}_{r_1}p_{i'r_2}(\varepsilon)\frac{{}_{r_1}p_{r_2j'}(\varepsilon)}{1 - {}_{r_1}p_{r_2r_2}(\varepsilon)} \\
&= p_{i'j'}(\varepsilon) + p_{i'r_1}(\varepsilon)\frac{p_{r_1j'}(\varepsilon)}{1 - p_{r_1r_1}(\varepsilon)} \\
&\quad + \frac{(p_{i'r_2}(\varepsilon) + p_{i'r_1}(\varepsilon)\frac{p_{r_1r_2}(\varepsilon)}{1-p_{r_1r_1}(\varepsilon)})(p_{r_2j'}(\varepsilon) + p_{r_2r_1}(\varepsilon)\frac{p_{r_1j'}(\varepsilon)}{1-p_{r_1r_1}(\varepsilon)})}{1 - p_{r_2r_2}(\varepsilon) - p_{r_2r_1}(\varepsilon)\frac{p_{r_1r_2}(\varepsilon)}{1-p_{r_1r_1}(\varepsilon)}} \\
&= p_{i'j'}(\varepsilon) + \frac{p_{i'r_1}(\varepsilon)p_{r_1j'}(\varepsilon)(1 - p_{r_2r_2}(\varepsilon)) + p_{i'r_1}(\varepsilon)p_{r_1r_2}(\varepsilon)p_{r_2j'}(\varepsilon)}{(1 - p_{r_1r_1}(\varepsilon))(1 - p_{r_2r_2}(\varepsilon)) - p_{r_1r_2}(\varepsilon)p_{r_2r_1}(\varepsilon)} \\
&\quad + \frac{p_{i'r_2}(\varepsilon)p_{r_2j'}(\varepsilon)(1 - p_{r_1r_1}(\varepsilon)) + p_{i'r_2}(\varepsilon)p_{r_2r_1}(\varepsilon)p_{r_1j'}(\varepsilon)}{(1 - p_{r_1r_1}(\varepsilon))(1 - p_{r_2r_2}(\varepsilon)) - p_{r_1r_2}(\varepsilon)p_{r_2r_1}(\varepsilon)} \\
&= {}_{r_2}p_{i'j'}(\varepsilon) + {}_{r_2}p_{i'r_1}(\varepsilon)\frac{{}_{r_2}p_{r_1j'}(\varepsilon)}{1 - {}_{r_2}p_{r_1r_1}(\varepsilon)} = {}_{\bar{r}'_{i,j,2}}p_{i'j'}(\varepsilon).
\end{aligned}
\tag{3.53}
$$

Therefore, by Lemma 2.4, the Laurent asymptotic expansions for transition probabilities ${}_{\bar{r}_{i,j,2}}p_{i'j'}(\varepsilon)$ and ${}_{\bar{r}'_{i,j,2}}p_{i'j'}(\varepsilon)$, given by the recurrent algorithm of sequential phase space reduction described above, are identical.

The proof of identity for Laurent asymptotic expansions of moments ${}_{\bar{r}_{i,j,2}}e_{i'j'}(k,\varepsilon)$ and ${}_{\bar{r}'_{i,j,2}}e_{i'j'}(k,\varepsilon)$, given by the recurrent algorithm of sequential phase space reduction described above, is analogous.

Let us take $n = N-2$. In this case, the semi-Markov process ${}_{\bar{r}_{i,j,N-2}}\eta_\varepsilon(t)$ has the phase space ${}_{\bar{r}_{i,j,N-2}}\mathbb{X} = \mathbb{X}\setminus\{r_{i,j,1}, r_{i,j,2},\ldots,r_{i,j,N-2}\} = \mathbb{X}_{ij} = \{i,j\}$, which is a two-state set.

By Theorem 3.1, the moment of hitting time $E_{i'j'}(k,\varepsilon)$ coincides for the initial semi-Markov processes $\eta_\varepsilon(t)$ and the reduced semi-Markov process ${}_{\bar{r}_{i,j,N-2}}\eta_\varepsilon(t)$, for every $k = 1,\ldots,d, i', j' \in \mathbb{X}_{ij}$. This obviously implies that these expectations are also invariant to any permutation $\bar{r}'_{i,j,N-2}$ of sequence $\bar{r}_{i,j,N-2}$.

Let us now exclude state i from the phase space of the semi-Markov process ${}_{\bar{r}_{i,j,N-2}}\eta_\varepsilon(t)$. The resulting semi-Markov process ${}_{\bar{r}_{i,j,N-1}}\eta_\varepsilon(t)$ can be considered as the semi-Markov process with transition period and the phase space $\mathbb{X}_{ij}$.

In this case, however, transition probabilities of the corresponding embedded Markov chain are concentrated on the one-point phase space $\mathbb{X}_{ij}\setminus\{i\} = \{j\}$, i.e., ${}_{\bar{r}_{i,j,N-1}}p_{i'j}(\varepsilon) = 1, i' \in \mathbb{X}_{ij}$.

As far as moments ${}_{\bar{r}_{i,j,N-1}}e_{i'j}(k,\varepsilon), k = 1,\ldots,d, i' \in \mathbb{X}_{ij}$ are concerned they can be computed using formulas given in relation (3.38), where roles of state r, the phase space $\mathbb{X}$, probabilities $p_{i'j}(\varepsilon), i', j' \in \mathbb{X}$, and moments $e_{i',j'}(k,\varepsilon), k = 1,\ldots,d, i', j' \in \mathbb{X}$ are played, respectively, by state i, space $\mathbb{X}_{ij}$, probabilities ${}_{\bar{r}_{i,j,N-2}}p_{i'j}(\varepsilon) = 1, i' \in \mathbb{X}_{ij}$, and moments ${}_{\bar{r}_{i,j,N-2}}e_{i'j}(k,\varepsilon), k = 1,\ldots,d, i' \in \mathbb{X}_{ij}$.

By Theorem 3.1, the moment of hitting time $E_{i'j}(k,\varepsilon)$ coincides for the initial semi-Markov processes $\eta_\varepsilon(t)$ and the reduced semi-Markov process ${}_{\bar{r}_{i,j,N-1}}\eta_\varepsilon(t)$, for every $k = 1,\ldots,d, i' \in \mathbb{X}_{ij}$. In the latter case, this moment coincides with the moment ${}_{\bar{r}_{i,j,N-1}}e_{i'j}(k,\varepsilon)$, since process ${}_{\bar{r}_{i,j,N-1}}\eta_\varepsilon(t)$ returns in state j after every jump. Thus,

$$E_{i'j}(k,\varepsilon) = {}_{\bar{r}_{i,j,N-1}}e_{i'j}(k,\varepsilon),\ k = 1,\ldots,d, i' \in \mathbb{X}_{ij}. \tag{3.54}$$

The algorithm described in Theorem 3.3 can be applied for constructing Laurent asymptotic expansions for moments $E_{i'j}(k,\varepsilon), k = 1,\ldots,d, i' \in \mathbb{X}_{ij}$,

$$E_{i'j}(k,\varepsilon) = \sum_{l=M^-_{i'j}[k]}^{M^+_{i'j}[k]} B_{i'j}[k,l]\varepsilon^l + \ddot{o}_{k,i'j}(\varepsilon^{M^+_{i'j}[k]}),\ \varepsilon \in (0,\varepsilon_0], \tag{3.55}$$

where (a) $M^\pm_{i'j}[k] = {}_{\bar{r}_{i,j,N-1}}m^\pm_{i'j}[k]$; (b) $B_{i'j}[k,l] = {}_{\bar{r}_{i,j,N-1}}b_{i'j}[k,l], l = M^-_{i'j}[k],\ldots,M^+_{i'j}[k]$; (c) $\ddot{o}_{k,i'j}(\varepsilon^{M^+_{i'j}[k]}) = {}_{\bar{r}_{i,j,N-1}}\dot{o}_{k,i'j}(\varepsilon^{{}_{\bar{r}_{i,j,N-1}}m^+_{i'j}[k]})$.

By the above remarks, the asymptotic expansion given in relation (3.55) is invariant with respect to the choice of sequence $\bar{r}_{i,j,N-2} = \langle r_{i,j,2},\ldots,r_{i,j,N-2}\rangle$. This legitimates notations (with omitted index ${}_{\bar{r}_{i,j,N-2}}$) used for parameters, coefficients, and remainder in the above asymptotic expansion.

The algorithm for construction of the Laurent asymptotic expansion for moments $E_{i'j}(k,\varepsilon), k = 1,\dots,d, i' \in \mathbb{X}_{ij}$, given in relation (3.55), can be repeated for every $i, j \in \mathbb{X}, i \neq j$. □

3.3.3 Asymptotic Expansions with Explicit Upper Bounds for Remainders for Transition Characteristics of Reduced Semi-Markov Processes

Relation (3.46) let us construct an algorithm for getting asymptotic expansions with explicit upper bounds for remainders, for non-absorption probabilities $\bar{p}_{rr}(\varepsilon)$.

Lemma 3.5. *Let conditions* **A** *and* **D′** *hold. Then, for every $r \in \mathbb{X}$, the pivotal $(\bar{l}^-_{rr}, \bar{l}^+_{rr})$-expansion for the non-absorption probability $\bar{p}_{rr}(\varepsilon)$ given in Lemma 3.4 is, also, a $(\bar{l}^-_{rr}, \bar{l}^+_{rr}, \bar{\delta}_{rr}, \bar{G}_{rr}, \bar{\varepsilon}_{rr})$-expansion, with parameters $\bar{\delta}_{rr}, \bar{G}_{rr}$, and $\bar{\varepsilon}_{rr}$, which can be computed according to the algorithm described below, in the proof of the lemma.*

Proof. Let $r \in \mathbb{Y}_r$.

First, propositions **(i)** of Lemmas 2.3 and 2.7 (the multiple summation rule) should be applied to the sum $\sum_{j \in \mathbb{Y}^+_{rr}} p_{rj}(\varepsilon)$.

Second, propositions **(i)** (the multiplication by constant −1) and **(ii)** (the summation with constant 1) of Lemmas 2.2 and 2.6 should be applied to the asymptotic expansion for probability $p_{rr}(\varepsilon)$ given in condition **D′**, in order to get the asymptotic expansion for function $1 - p_{rr}(\varepsilon)$.

Third, Lemmas 2.1 and 2.5 should be applied to the asymptotic expansion for function $\bar{p}_{rr}(\varepsilon)$ given in two alternative forms by relation (3.46).

This yields the corresponding pivotal the $(\bar{l}^-_{rr}, \bar{l}^+_{rr})$-expansion for probabilities $\bar{p}_{rr}(\varepsilon)$, given in Lemma 3.4, and proves that this expansion is a $(\bar{l}^-_{rr}, \bar{l}^+_{rr}, \bar{\delta}_{rr}, \bar{G}_{rr}, \bar{\varepsilon}_{rr})$-expansion, with parameters computed in the process of realization of the above algorithm. The case $r \notin \mathbb{Y}_r$ is trivial, since, in this case, probability $\bar{p}_{rr}(\varepsilon) \equiv 1$. □

Relation (3.31) let us construct an algorithm for getting asymptotic expansions with explicit upper bounds for remainders, for transition probabilities ${}_r p_{ij}(\varepsilon)$.

Let us introduce parameter,

$$\delta^\circ = \min_{j \in \mathbb{Y}_i, i \in \mathbb{X}} \delta_{ij}. \tag{3.56}$$

Obviously, inequalities $\delta_{ij} \geq \delta^\circ, j \in \mathbb{Y}_i, i \in \mathbb{X}$ hold for parameters δ_{ij} appearing in upper bounds for the remainders of asymptotic expansions in condition **D′**.

Theorem 3.5. *Conditions* **A** *and* **D′**, *assumed to hold for the Markov chains $\eta_{\varepsilon,n}$, also hold for the reduced Markov chains ${}_r\eta_{\varepsilon,n}$, for every $r \in \mathbb{X}$. Also, for every $j \in {}_r\mathbb{Y}_i$, $i \in {}_r\mathbb{X}$, $r \in \mathbb{X}$, the pivotal $({}_r l^-_{ij}, {}_r l^+_{ij})$-expansion for the transition probability ${}_r p_{ij}(\varepsilon)$ given in Theorem 3.2 is a $({}_r l^-_{ij}, {}_r l^+_{ij}, {}_r\delta_{ij}, {}_r G_{ij}, {}_r\varepsilon_{ij})$-expansion appearing in condition* **D′** *for the Markov chains ${}_r\eta_{\varepsilon,n}$. Parameters ${}_r\delta_{ij}$, ${}_r G_{ij}$, and ${}_r\varepsilon_{ij}$ can be computed using the algorithm described below, in the proof of the theorem. The inequalities ${}_r\delta_{ij} \geq \delta^\circ$, $j \in {}_r\mathbb{Y}_i$, $i \in {}_r\mathbb{X}$, $r \in \mathbb{X}$ hold.*

Proof. By Lemma 3.2, condition **A** holds for the Markov chains ${}_r\eta_{\varepsilon,n}$, with the same parameter ε_0 as for the Markov chains $\eta_{\varepsilon,n}$ and with the transition sets ${}_r\mathbb{Y}_i, i \in {}_r\mathbb{X}$ given by relation (3.42).

Let us prove that condition $\mathbf{D}'$ holds for the Markov chains ${}_r\eta_{\varepsilon,n}$, with the same parameter ε_0 as for the Markov chains $\eta_{\varepsilon,n}$ and the transition sets ${}_r\mathbb{Y}_i, i \in {}_r\mathbb{X}$ given by relation (3.42). Let $j, r \in \mathbb{Y}_i \cap \mathbb{Y}_r$.

First, propositions **(v)** (the division rule) of Lemmas 2.2 and 2.6 should be applied to the quotient $\frac{p_{rj}(\varepsilon)}{1-p_{rr}(\varepsilon)}$.

Second, propositions **(iii)** (the multiplication rule) of Lemmas 2.2 and 2.6 should be applied to the product $p_{ir}(\varepsilon) \cdot \frac{p_{rj}(\varepsilon)}{1-p_{rr}(\varepsilon)}$.

Third, propositions **(ii)** (the summation rule) of Lemmas 2.2 and 2.6 should be applied to sum ${}_rp_{ij}(\varepsilon) = p_{ij}(\varepsilon) + p_{ir}(\varepsilon) \cdot \frac{p_{rj}(\varepsilon)}{1-p_{rr}(\varepsilon)}$.

The asymptotic expansions for probabilities $p_{ir}(\varepsilon), p_{rj}(\varepsilon)$, and $p_{ij}(\varepsilon)$, given in condition $\mathbf{D}'$, and probability $1 - p_{rr}(\varepsilon)$, given in Lemmas 3.4 and 3.5, should be used.

This yields the corresponding pivotal $({}_rl^-_{ij}, {}_rl^+_{ij})$-expansions for transition probabilities ${}_rp_{ij}(\varepsilon), j \in {}_r\mathbb{Y}_i, i \in {}_r\mathbb{X}, r \in \mathbb{X}$, given in Theorem 3.2, and proves that these expansions are $({}_rl^-_{ij}, {}_rl^+_{ij}, {}_r\delta_{ij}, {}_rG_{ij}, {}_r\varepsilon_{ij})$-expansions, with parameters computed in the process of realization of the above algorithm.

If $j \notin \mathbb{Y}_i$, then $p_{ij}(\varepsilon) \equiv 0$; if $j \notin \mathbb{Y}_r$, then $p_{rj}(\varepsilon) \equiv 0$; if $r \notin \mathbb{Y}_i$, then $p_{ir}(\varepsilon) \equiv 0$; if $r \notin \mathbb{Y}_r$, then $1 - p_{rr}(\varepsilon) \equiv 1$. In these cases, the above algorithm is readily simplified. Thus, condition $\mathbf{D}'$ holds for the reduced Markov chains ${}_r\eta_{\varepsilon,n}$.

Inequalities ${}_r\delta_{ij} \geq \delta^\circ, j \in {}_r\mathbb{Y}_i, i \in {}_r\mathbb{X}, r \in \mathbb{X}$ hold, by proposition **(iii)** of Lemma 2.8. □

Relation (3.38) let us construct an algorithm for getting asymptotic expansions with explicit upper bounds for remainders, for moments probabilities ${}_re_{ij}(k,\varepsilon)$.

Let us introduce parameters,

$$\delta^*[k] = \min_{j \in \mathbb{Y}_i, i \in \mathbb{X}} (\delta_{ij} \wedge \dot{\delta}_{ij}[1] \wedge \cdots \wedge \dot{\delta}_{ij}[k]), \ k = 1, \ldots, d. \tag{3.57}$$

Obviously, inequalities $\delta_{ij}, \dot{\delta}_{ij}[1], \ldots, \dot{\delta}_{ij}[k] \geq \delta^*[k], k = 1, \ldots, d, j \in \mathbb{Y}_i, i \in \mathbb{X}$ hold for parameters δ_{ij} and $\dot{\delta}_{ij}[1], \ldots, \dot{\delta}_{ij}[d]$ appearing in upper bounds for the remainders of asymptotic expansions, respectively, in conditions $\mathbf{D}'$ and $\mathbf{E}'_d$. □

Remark 3.5. The algorithm described in the proof of Theorem 3.5 also yields the pivotal asymptotic expansions with explicit upper bound for remainders for the transition probabilities ${}_rp_{rj}(\varepsilon), j \in {}_r\mathbb{Y}_r, r \in \mathbb{X}$. They, actually, are obtained at the first step of the algorithm presented in the above proof.

Theorem 3.6. *Conditions* **A**, **B**, $\mathbf{C}_d$, $\mathbf{D}'$, *and* $\mathbf{E}'_d$, *assumed to hold for the semi-Markov processes* $\eta_\varepsilon(t)$, *also hold for the reduced semi-Markov processes* ${}_r\eta_\varepsilon(t)$, *for every* $r \in \mathbb{X}$. *Also, for every* $k = 1, \ldots, d, j \in {}_r\mathbb{Y}_i, i \in {}_r\mathbb{X}, r \in \mathbb{X}$, *the pivotal* $({}_rm^-_{ij}[k], {}_rm^+_{ij}[k])$*-expansion for moment* ${}_re_{ij}(k,\varepsilon)$ *given in Theorem 3.3 is a*

$({}_rm^-_{ij}[k], {}_rm^+_{ij}[k], {}_r\dot{\delta}_{ij}[k], {}_r\dot{G}_{ij}[k], {}_r\dot{\varepsilon}_{ij}[k])$-expansion appearing in condition $\mathbf{E}'_d$ for the semi-Markov processes ${}_r\eta_\varepsilon(t)$. Parameters ${}_r\dot{\delta}_{ij}[k]$, ${}_r\dot{G}_{ij}[k]$, and ${}_r\dot{\varepsilon}_{ij}[k]$ can be computed using the recurrent algorithm described below, in the proof of the theorem. Inequalities ${}_r\dot{\delta}_{ij}[k] \geq \delta^*[k], k = 1,\ldots,d, j \in {}_r\mathbb{Y}_i, i \in {}_r\mathbb{X}, r \in \mathbb{X}$ hold.

Proof. Conditions **A** and **D**$'$ hold for the semi-Markov processes ${}_r\eta_\varepsilon(t)$, respectively, by Lemma 3.2 and Theorem 3.5, with the same parameter ε_0 as for the semi-Markov processes $\eta_\varepsilon(t)$, and the transition sets ${}_r\mathbb{Y}_i, i \in {}_r\mathbb{X}$ given by relation (3.42). Also conditions **B** and $\mathbf{C}_d$ hold for processes ${}_r\eta_\varepsilon(t)$, by Lemma 3.3.

Let us prove that condition $\mathbf{E}'_d$ holds for the semi-Markov processes ${}_r\eta_\varepsilon(t)$, with the same parameter ε_0 and the transition sets ${}_r\mathbb{Y}_i, i \in {}_r\mathbb{X}$ given by relation (3.42).

Let $j, r \in \mathbb{Y}_i \cap \mathbb{Y}_r$.

At the first recurrent step, we construct the asymptotic expansions with explicit upper bounds for remainders for expectations ${}_re_{rj}(1,\varepsilon)$ and ${}_re_{ij}(1,\varepsilon)$ using formulas (3.39) and the corresponding asymptotic expansions appearing in conditions **D**$'$ and $\mathbf{E}'_d$,

First, propositions **(iii)** (the multiplication rule) of Lemmas 2.2 and 2.6 should be applied to the product $e_{rr}(1,\varepsilon) \cdot {}_re_{rj}(0,\varepsilon)$. Here, the asymptotic expansion for probability ${}_re_{rj}(0,\varepsilon) = {}_rp_{rj}(\varepsilon)$ constructed in Theorem 3.5 should be used.

Second, propositions **(ii)** (the summation rule) of Lemmas 2.2 and 2.6 should be applied to the sum $e_{rj}(1,\varepsilon) + e_{rr}(1,\varepsilon) \cdot {}_re_{rj}(0,\varepsilon)$.

Third, propositions **(v)** (the division rule) of Lemmas 2.2 and 2.6 should be applied to the quotient ${}_re_{rj}(1,\varepsilon) = \frac{1}{1-p_{rr}(\varepsilon)}(e_{rj}(1,\varepsilon) + e_{rr}(1,\varepsilon) \cdot {}_re_{rj}(0,\varepsilon))$. Here, the asymptotic expansion for probability $1 - p_{rr}(\varepsilon)$ constructed in Lemmas 3.4 and 3.5 should be used.

Fourth, propositions **(iii)** (the multiplication rule) of Lemmas 2.2 and 2.6 should be applied to the products $e_{ir}(1,\varepsilon) \cdot {}_re_{rj}(0,\varepsilon)$ and $p_{ir}(\varepsilon) \cdot {}_re_{rj}(1,\varepsilon)$. Here, the asymptotic expansion for expectations ${}_re_{rj}(0,\varepsilon) = {}_rp_{rj}(\varepsilon)$ constructed in Theorem 3.6 and ${}_re_{rj}(1,\varepsilon)$ constructed above should be used.

Fifth, the propositions **(i)** (the multiple summation rule) of Lemmas 2.3 and 2.7 should be applied to sum ${}_re_{ij}(1,\varepsilon) = e_{ij}(1,\varepsilon) + e_{ir}(1,\varepsilon) \cdot {}_re_{rj}(0,\varepsilon) + p_{ir}(\varepsilon) \cdot {}_re_{rj}(1,\varepsilon)$.

The asymptotic expansions for probabilities $p_{ij}(\varepsilon), p_{ir}(\varepsilon)$, and $p_{rj}(\varepsilon)$, given in condition **D**$'$ and expectations $e_{ij}(1,\varepsilon), e_{ir}(1,\varepsilon), e_{rr}(1,\varepsilon)$, and $e_{rj}(1,\varepsilon)$, given in condition $\mathbf{E}'_d$ are used in the first recurrent step of the algorithm described above.

If $j \notin \mathbb{Y}_i$, then $p_{ij}(\varepsilon) \equiv 0$ and $e_{ij}(1,\varepsilon) \equiv 0$; if $j \notin \mathbb{Y}_r$, then $p_{rj}(\varepsilon) \equiv 0$ and $e_{rj}(1,\varepsilon) \equiv 0$; if $r \notin \mathbb{Y}_i$, then $p_{ir}(\varepsilon) \equiv 0$ and $e_{ir}(1,\varepsilon) \equiv 0$; if $r \notin \mathbb{Y}_r$, then $1 - p_{rr}(\varepsilon) \equiv 1$ and $e_{rr}(1,\varepsilon) \equiv 0$. In these cases, the above algorithm is readily simplified.

The first recurrent step of the algorithm described above yields the corresponding pivotal $({}_rm^-_{ij}[1], {}_rm^+_{ij}[1])$-expansions for expectations ${}_re_{ij}(1,\varepsilon), j \in {}_r\mathbb{Y}_i, i \in {}_r\mathbb{X}, r \in \mathbb{X}$, given in Theorem 3.3, and proves that these expansions are $({}_rm^-_{ij}[1], {}_rm^+_{ij}[1], {}_r\dot{\delta}_{ij}[1], {}_r\dot{G}_{ij}[1], {}_r\dot{\varepsilon}_{ij}[1])$-expansions, with parameters computed in the process of realization of the above algorithm.

The algorithm described above also yields the pivotal asymptotic expansions for expectations ${}_re_{rj}(1,\varepsilon), j \in {}_r\mathbb{Y}_r, r \in \mathbb{X}$. They, actually, are obtained at the third step of the above algorithm.

Five steps of the algorithm described above should be recurrently repeated for $k = 2, \ldots, d$.

Let us assume that the corresponding asymptotic expansions with explicit upper bounds for remainders for moments ${}_re_{rj}(l, \varepsilon)$, ${}_re_{ij}(l, \varepsilon), l = 1, \ldots, k-1$ have been already constructed with the use of formulas (3.38) and the corresponding asymptotic expansions appearing in conditions $\mathbf{D}'$ and $\mathbf{E}'_d$. In this case, the asymptotic expansions with explicit upper bounds for remainders for expectations ${}_re_{rj}(k, \varepsilon)$, ${}_re_{ij}(k, \varepsilon)$ can be constructed using the above expansions, formulas (3.38), and the corresponding asymptotic expansions appearing in conditions $\mathbf{D}'$ and $\mathbf{E}'_d$, in the following way.

First, propositions **(i)** (the multiplication by constant rule) and **(iii)** (the multiplication rule) of Lemmas 2.2 and 2.6 should be applied to the products $\binom{k}{l} e_{ir}(k-l, \varepsilon) \cdot {}_re_{rj}(l, \varepsilon), l = 0, \ldots, k-1$.

Second, propositions **(i)** (the multiple summation rule) of Lemmas 2.3 and 2.7 should be applied to the sum $e_{rj}(k, \varepsilon) + \sum_{l=0}^{k-1} \binom{k}{l} e_{rr}(k-l, \varepsilon) \cdot {}_re_{rj}(l, \varepsilon)$

Third, propositions **(v)** (the division rule) of Lemmas 2.2 and 2.6 should be applied to the quotient ${}_re_{rj}(k, \varepsilon) = \frac{1}{1 - p_{rr}(\varepsilon)} (e_{rj}(k, \varepsilon) + \sum_{l=0}^{k-1} \binom{k}{l} e_{rr}(k-l, \varepsilon) \cdot {}_re_{rj}(l, \varepsilon))$.

Fourth, propositions **(i)** (the multiplication by constant rule) and **(iii)** (the multiplication rule) of Lemmas 2.2 and 2.6 should be applied to the products $\binom{k}{l} e_{ir}(k - l, \varepsilon) \cdot {}_re_{rj}(l, \varepsilon), l = 0, k-1$ and $p_{ir}(\varepsilon) \cdot {}_re_{rj}(k, \varepsilon)$. Here, the corresponding asymptotic expansion with the explicit upper bound for remainder for moment ${}_re_{rj}(k, \varepsilon)$, constructed above, should be used.

Fifth, the propositions **(i)** (the multiple summation rule) of Lemmas 2.3 and 2.7 should be applied to sum ${}_re_{ij}(k, \varepsilon) = e_{ij}(k, \varepsilon) + \sum_{l=0}^{k-1} \binom{k}{l} e_{ir}(k-l, \varepsilon) \cdot {}_re_{rj}(l, \varepsilon) + p_{ir}(\varepsilon) \cdot {}_re_{rj}(k, \varepsilon)$.

The asymptotic expansions for probabilities $p_{ij}(\varepsilon), p_{ir}(\varepsilon)$, and $p_{rj}(\varepsilon)$, given in condition $\mathbf{D}'$ and expectations $e_{ij}(l, \varepsilon), e_{ir}(l, \varepsilon), e_{rr}(l, \varepsilon), e_{rj}(l, \varepsilon), l = 1, \ldots, k$, given in condition $\mathbf{E}'_d$, are used at the k-th recurrent step of the algorithm.

If $j \notin \mathbb{Y}_i$, then $p_{ij}(\varepsilon) \equiv 0$ and $e_{ij}(k, \varepsilon) \equiv 0, k = 1, \ldots, d$; if $j \notin \mathbb{Y}_r$, then $p_{rj}(\varepsilon) \equiv 0$ and $e_{rj}(k, \varepsilon) \equiv 0, k = 1, \ldots, d$; if $r \notin \mathbb{Y}_i$, then $p_{ir}(\varepsilon) \equiv 0$ and $e_{ir}(k, \varepsilon) \equiv 0, k = 1, \ldots, d$; if $r \notin \mathbb{Y}_r$, then $1 - p_{rr}(\varepsilon) \equiv 1$ and $e_{rr}(k, \varepsilon) \equiv 0, k = 1, \ldots, d$. In these cases, the above recurrent algorithm is readily simplified.

The k-th recurrent step of the algorithm described above yields the corresponding pivotal $({}_rm_{ij}^-[k], {}_rm_{ij}^+[k])$-expansions for expectations ${}_re_{ij}(k, \varepsilon), j \in {}_r\mathbb{Y}_i, i \in {}_r\mathbb{X}, r \in \mathbb{X}$, given in Theorem 3.3, and proves that these expansions are $({}_rm_{ij}^-[k], {}_rm_{ij}^+[k], {}_r\dot{\delta}_{ij}[k], {}_r\dot{G}_{ij}[k], {}_r\dot{\varepsilon}_{ij}[k])$-expansions, with parameters computed in the process of realization of the above algorithm.

As it was already mentioned above, five steps of the above algorithm should be recurrently repeated for $k = 1, 2, \ldots, d$.

Thus, condition $\mathbf{E}'_d$ holds for the reduced semi-Markov processes ${}_r\eta_\varepsilon(t)$.

Inequalities ${}_r\dot{\delta}_{ij}[k] \geq \delta^*[k], k = 1, \ldots, d, j \in {}_r\mathbb{Y}_i, i \in {}_r\mathbb{X}, r \in \mathbb{X}$ hold, by proposition **(iii)** of Lemma 2.8. □

Remark 3.6. The algorithm described in the proof of Theorem 3.6 also yields the pivotal asymptotic expansion for moments ${}_re_{rj}(k, \varepsilon), k = 1, \ldots, d$.

It is worth to note that, despite bulky forms, recurrent formulas for parameters of upper bounds for remainders, in the asymptotic expansions given in Lemma 3.5 and Theorems 3.5 and 3.6, are computationally effective.

3.3.4 Asymptotic Expansions with Explicit Upper Bounds for Remainders for Moments of Hitting Times for Reduced Semi-Markov Processes

Let us again, $i, j \in \mathbb{X}, i \neq j$ and let $\bar{r}_{i,j,N} = \langle r_{i,j,1}, \ldots, r_{i,j,N} \rangle = \langle r_{i,1}, \ldots, r_{i,N-2}, i, j \rangle$ be a permutation of the sequence $\langle 1, \ldots, N \rangle$ such that $r_{i,j,N-1} = i, r_{i,j,N} = j$, and let $\bar{r}_{i,j,n} = \langle r_{i,j,1}, \ldots, r_{i,j,n} \rangle$, $n = 1, \ldots, N$ be the corresponding chain of growing sequences of states from space $\mathbb{X}$.

Theorem 3.7. *Let conditions* **A**, **B**, $\mathbf{C}_d$, $\mathbf{D}'$, *and* $\mathbf{E}'_d$ *hold for the semi-Markov processes* $\eta_\varepsilon(t)$. *Then, for every* $i, j \in \mathbb{X}, i \neq j$, *the pivotal* $(M^-_{i'j}[k], M^+_{i'j}[k])$*-expansions for moments of hitting time* $E_{i'j}(k, \varepsilon), k = 1, \ldots, d, i' = i, j$ *given in Theorem 3.4 and obtained as the result of sequential exclusion of states* $r_{i,1}, \ldots, r_{i,N-1}, i$ *from the phase space* $\mathbb{X}$ *of the processes* $\eta_\varepsilon(t)$, *are* $(M^-_{i'j}[k], M^+_{i'j}[k], {}_{\bar{r}_{i,j,N-1}}\dot{\delta}_{i'j}[k], {}_{\bar{r}_{i,j,N-1}}\dot{G}_{i'j}[k], {}_{\bar{r}_{i,j,N-1}}\dot{\varepsilon}_{i'j}[k])$*-expansions. Parameters* ${}_{\bar{r}_{i,j,N-1}}\dot{\delta}_{i'j}[k]$, ${}_{\bar{r}_{i,j,N-1}}\dot{G}_{i'j}[k]$, *and* ${}_{\bar{r}_{i,j,N-1}}\dot{\varepsilon}_{i'j}[k]$ *can be computed using the algorithm described below, in the proof of the theorem.*

Proof. Let us assume that $p_{\varepsilon,i} + p_{\varepsilon,j} = 1$.

Let us repeat the procedure of sequential exclusion of states $r_1, \ldots, r_n$ from the phase space $\mathbb{X}$ of the semi-Markov process $\eta_\varepsilon(t)$ described in the proof of Theorem 3.4. As the result we construct the standard reduced semi-Markov process ${}_{\bar{r}_{i,j,n}}\eta_\varepsilon(t)$ with the phase space ${}_{\bar{r}_{i,j,n}}\mathbb{X} = \mathbb{X} \setminus \{r_{i,j,1}, r_{i,j,2}, \ldots, r_{i,j,n}\}$.

The transition probabilities of the embedded Markov chain ${}_{\bar{r}_{i,j,n}}p_{i'j'}(\varepsilon)$, $i', j' \in {}_{\bar{r}_{i,j,n}}\mathbb{X}$, and the moments of transition times ${}_{\bar{r}_{i,j,n}}e_{i'j'}(k, \varepsilon), k = 1, \ldots, d$, $i', j' \in {}_{\bar{r}_{i,n}}\mathbb{X}$ are determined for the semi-Markov process ${}_{\bar{r}_{i,j,n}}\eta_\varepsilon(t)$ by the transition probabilities and the corresponding moments of transition times for the process ${}_{\bar{r}_{i,j,n-1}}\eta_\varepsilon(t)$, respectively, via relations (3.31) and (3.38).

Relations (3.31) and (3.38) also give formulas for probabilities ${}_{\bar{r}_{i,j,n}}p_{r_{i,j,n-1}j'}(\varepsilon)$, $j' \in {}_{\bar{r}_{i,j,n}}\mathbb{X}$ and moments ${}_{\bar{r}_{i,j,n}}e_{r_{i,j,n-1}j'}(k, \varepsilon), k = 1, \ldots, d$, $j' \in {}_{\bar{r}_{i,j,n}}\mathbb{X}$ determined by the transition probabilities and the corresponding moments of sojourn times for the process ${}_{\bar{r}_{i,j,n-1}}\eta_\varepsilon(t)$.

By Theorem 3.1, the moment of hitting time $E_{i'j'}(k, \varepsilon)$ coincides for the semi-Markov processes ${}_{\bar{r}_{i,j,0}}\eta_\varepsilon(t)$, ${}_{\bar{r}_{i,j,1}}\eta_\varepsilon(t), \ldots$, ${}_{\bar{r}_{i,j,n}}\eta_\varepsilon(t)$, for every $k = 1, \ldots, d, i', j' \in {}_{\bar{r}_{i,j,n}}\mathbb{X}$.

By Theorems 3.2, 3.3, 3.5, and 3.6 the semi-Markov processes ${}_{\bar{r}_{i,j,n}}\eta_\varepsilon(t)$ satisfies conditions **B**, $\mathbf{C}_d$ and, also, conditions **A**, $\mathbf{D}'$, and $\mathbf{E}'_d$. The transition sets ${}_{\bar{r}_{i,j,n}}\mathbb{Y}_{i'}, i' \in {}_{\bar{r}_{i,j,n}}\mathbb{X}$ determined by the transition sets ${}_{\bar{r}_{i,j,n-1}}\mathbb{Y}_{i'}, i' \in {}_{\bar{r}_{i,j,n-1}}\mathbb{X}$, via relation (3.42) given in Lemma 3.2.

For every $j' \in {}_{\bar{r}_{i,j,n}}\mathbb{Y}_{i'}, i' \in {}_{\bar{r}_{i,j,n}}\mathbb{X}$, the pivotal $({}_{\bar{r}_{i,j,n}}l^-_{i'j'}, {}_{\bar{r}_{i,j,n}}l^+_{i'j'})$-expansion for transition probability ${}_{\bar{r}_{i,j,n}}p_{i'j'}(\varepsilon)$, given in Theorem 3.2, is, by Theorem 3.5, a $({}_{\bar{r}_{i,j,n}}l^-_{i'j'}, {}_{\bar{r}_{i,j,n}}l^+_{i'j'}, {}_{\bar{r}_{i,j,n}}\delta_{i'j'}, {}_{\bar{r}_{i,j,n}}G_{i'j'}, {}_{\bar{r}_{i,j,n}}\varepsilon_{i'j'})$-expansion, with parameters ${}_{\bar{r}_{i,j,n}}\delta_{i'j'}$, ${}_{\bar{r}_{i,j,n}}G_{i'j'}$ and ${}_{\bar{r}_{i,j,n}}\varepsilon_{i'j'}$ given in this theorem. Analogously, for every $j' \in {}_{\bar{r}_{i,j,n}}\mathbb{Y}_{i'}, i' \in {}_{\bar{r}_{i,j,n}}\mathbb{X}$, $k = 1,\ldots,d$, the pivotal $({}_{\bar{r}_{i,j,n}}m^-_{i'j'}[k], {}_{\bar{r}_{i,j,n}}m^+_{i'j'}[k])$-expansion for moment ${}_{\bar{r}_{i,j,n}}e_{i'j'}(k,\varepsilon)$, given in Theorem 3.3, is, by Theorem 3.6, a $({}_{\bar{r}_{i,j,n}}m^-_{i'j'}[k], {}_{\bar{r}_{i,j,n}}m^+_{i'j'}[k], {}_{\bar{r}_{i,j,n}}\dot{\delta}_{i'j'}[k], {}_{\bar{r}_{i,j,n}}\dot{G}_{i'j'}[k], {}_{\bar{r}_{i,j,n}}\dot{\varepsilon}_{i'j'}[k])$-expansion, with parameters ${}_{\bar{r}_{i,j,n}}\dot{\delta}_{i'j'}[k]$, ${}_{\bar{r}_{i,j,n}}\dot{G}_{i'j'}[k]$ and ${}_{\bar{r}_{i,j,n}}\dot{\varepsilon}_{i'j'}[k]$ given in this theorem. Also, by Theorem 3.6, the inequalities ${}_{\bar{r}_{i,j,n}}\dot{\delta}_{i'j'}[k] \geq \delta^*[k], j' \in {}_{\bar{r}_{i,j,n}}\mathbb{Y}_{i'}, i' \in {}_{\bar{r}_{i,j,n}}\mathbb{X}$ hold.

The algorithm described above has a recurrent form and it should be realized sequentially for the reduced semi-Markov processes ${}_{\bar{r}_{i,1}}\eta_\varepsilon(t), \ldots, {}_{\bar{r}_{i,n}}\eta_\varepsilon(t)$ starting from the initial semi-Markov process ${}_{\bar{r}_{i,0}}\eta_\varepsilon(t) = \eta_\varepsilon(t)$.

Let us take $n = N-2$. In this case, the semi-Markov process ${}_{\bar{r}_{i,j,N-2}}\eta_\varepsilon(t)$ has the phase space ${}_{\bar{r}_{i,j,N-2}}\mathbb{X} = \mathbb{X} \setminus \{r_{i,j,1}, r_{i,j,2}, \ldots, r_{i,j,N-2}\} = \mathbb{X}_{ij} = \{i, j\}$, which is a two-state set.

As in Theorem 3.4, let us now exclude state i from the phase space of the semi-Markov process ${}_{\bar{r}_{i,j,N-2}}\eta_\varepsilon(t)$. The resulting semi-Markov process ${}_{\bar{r}_{i,j,N-1}}\eta_\varepsilon(t)$ can be considered as the semi-Markov process with transition period and the phase space $\mathbb{X}_{ij}$. In this case, however, transition probabilities of the corresponding embedded Markov chain are concentrated on the one-point phase space $\mathbb{X}_{ij} \setminus \{i\} = \{j\}$, i.e., ${}_{\bar{r}_{i,j,N-1}}p_{i'j}(\varepsilon) = 1, i' \in \mathbb{X}_{ij}$.

As far as moments ${}_{\bar{r}_{i,j,N-1}}e_{i'j}(k,\varepsilon), k = 1,\ldots,d, i' \in \mathbb{X}_{ij}$ are concerned they can be computed using formulas given in relation (3.38), where roles of state r, the phase space $\mathbb{X}$, probabilities $p_{i'j}(\varepsilon), i', j \in \mathbb{X}$, and moments $e_{i',j'}(k,\varepsilon), k = 1,\ldots,d, i', j' \in \mathbb{X}$ are played, respectively, by state i, space $\mathbb{X}_{ij}$, probabilities ${}_{\bar{r}_{i,j,N-2}}p_{i'j}(\varepsilon) = 1, i' \in \mathbb{X}_{ij}$ and moments ${}_{\bar{r}_{i,j,N-2}}e_{i'j}(k,\varepsilon), k = 1,\ldots,d, i' \in \mathbb{X}_{ij}$.

By Theorem 3.1, the moment of hitting time $E_{i'j}(k,\varepsilon)$ coincides for the initial semi-Markov processes $\eta_\varepsilon(t)$ and the reduced semi-Markov process ${}_{\bar{r}_{i,j,N-1}}\eta_\varepsilon(t)$, for every $k = 1,\ldots,d, i', j \in \mathbb{X}_{ij}$. In the latter case, this moment coincides with the moment ${}_{\bar{r}_{i,j,N-1}}e_{i'j}(k,\varepsilon)$, since process ${}_{\bar{r}_{i,j,N-1}}\eta_\varepsilon(t)$ returns in state j after every jump. Thus,

$$E_{i'j}(k,\varepsilon) = {}_{\bar{r}_{i,j,N-1}}e_{i'j}(k,\varepsilon), \ k = 1,\ldots,d, i' \in \mathbb{X}_{ij}. \tag{3.58}$$

For every $k = 1,\ldots,d, i' \in \mathbb{X}_{ij}$, the pivotal $({}_{\bar{r}_{i,j,N-1}}m^-_{i'j}[k], {}_{\bar{r}_{i,j,n}}m^+_{i'j}[k])$-expansion for moment $E_{i'j}(k,\varepsilon) = {}_{\bar{r}_{i,j,N-1}}e_{i'j}(k,\varepsilon)$, given in Theorem 3.3, is, by Theorem 3.6, a $({}_{\bar{r}_{i,j,N-1}}m^-_{i'j}[k], {}_{\bar{r}_{i,j,N-1}}m^+_{i'j}[k], {}_{\bar{r}_{i,j,N-1}}\dot{\delta}_{i',j}[k], {}_{\bar{r}_{i,j,N-1}}\dot{G}_{i'j}[k], {}_{\bar{r}_{i,j,N-1}}\dot{\varepsilon}_{i'j}[k])$-expansion, with parameters $M^{\pm}_{i'j}[k] = {}_{\bar{r}_{i,j,N-1}}m^{\pm}_{i'j}[k]$, ${}_{\bar{r}_{i,j,N-1}}\dot{\delta}_{i'j}[k]$, ${}_{\bar{r}_{i,j,N-1}}\dot{G}_{i'j}[k]$ and ${}_{\bar{r}_{i,j,N-1}}\dot{\varepsilon}_{i'j}[k]$ given in this theorem. □

Remark 3.7. By Lemma 2.8 and Theorem 3.6, the inequalities ${}_{\bar{r}_{i,j,N-1}}\dot{\delta}_{i'j}[k] \geq \delta^*[k]$, $k = 1,\ldots,d, i' \in \mathbb{X}_{ij}$ hold. This makes it possible to rewrite functions $E_{i'j}(k,\varepsilon)$, $k = 1,\ldots,d, i' = i, j$ as pivotal $(M^-_{i'j}[k], M^+_{i'j}[k], \delta^*[k], {}_{\bar{r}_{i,j,N-1}}\dot{G}^*_{i'j}[k], {}_{\bar{r}_{i,j,N-1}}\dot{\varepsilon}_{i'j}[k])$-expansions, with parameters

$${}_{\bar{r}_{i,j,N-1}}\dot{G}^*_{i'j}[k] = {}_{\bar{r}_{i,j,N-1}}\dot{G}_{i'j}[k] \cdot ({}_{\bar{r}_{i,j,N-1}}\dot{\varepsilon}_{i'j}[k])^{({}_{\bar{r}_{i,j,N-1}}\dot{\delta}_{i'j}[k] - \delta^*[k])}.$$

Chapter 4
Asymptotic Expansions for Stationary Distributions of Nonlinearly Perturbed Semi-Markov Processes

In Chapter 4, new algorithms for construction of asymptotic expansions for stationary distributions of nonlinearly perturbed semi-Markov processes with finite phase spaces are presented. These algorithms are based on the technique of sequential phase space reduction, which can be applied to processes with an arbitrary asymptotic communicative structure of phase spaces. Asymptotic expansions are given in two forms, without and with explicit upper bounds for remainders.

We would like to make here comments analogous to those given in the preamble of Section 3.3.

An explicit writing down of the corresponding operational formulas representing the recurrent algorithms described below in Theorems 4.1–4.8 (which could be given, say, as corollaries of these lemmas and theorems) would, in fact, replicate the above proofs in the formal form, require implementation of a huge number of intermediate notations, take too much space, etc., but would not add any new essential information about the corresponding algorithms.

That is why the decision was made, just, to say in each theorem that the description of the corresponding algorithm is given in its proof. This makes formulations slightly unusual. But, as we think, this is the most compact way for presentation of the corresponding asymptotic results and algorithms.

4.1 Asymptotic Expansions with Remainders Given in the Standard Form for Stationary Distributions of Perturbed Semi-Markov Processes

In this section, we describe an algorithm for construction of asymptotic expansions with remainders given in the standard form for stationary distributions of nonlinearly perturbed semi-Markov processes.

D. Silvestrov, S. Silvestrov, *Nonlinearly Perturbed Semi-Markov Processes*,
SpringerBriefs in Probability and Mathematical Statistics,
DOI 10.1007/978-3-319-60988-1_4

4.1.1 Asymptotic Expansions with Remainders Given in the Standard Form for Expectations of Return Times

The algorithms presented below are based on the formula $\pi_i(\varepsilon) = \frac{e_i(\varepsilon)}{E_{ii}(1,\varepsilon)}$, $i \in \mathbb{X}$, which connect stationary probabilities $\pi_i(\varepsilon)$ of ergodic semi-Markov process with expectations of the corresponding return times $E_{ii}(1,\varepsilon)$. Laurent asymptotic expansions for expectations of return times $E_{ii}(1,\varepsilon)$ are constructed with the use of the time-space screening procedure and the algorithm of sequential phase space reduction for moments of hitting times described in Sections 3.2 and 3.3. A slight difference is that, in this case, the phase space reduction should be continued up to exclusion from the phase space $\mathbb{X}$ all states except i. Also, we need to construct Laurent asymptotic expansions only for expectations and, thus, the corresponding recurrent sub-procedures connected with construction of expansions for high order moments can be omitted.

In what follows, let $\bar{r}_{i,N} = \langle r_{i,1}, \ldots, r_{i,N} \rangle = \langle r_{i,1}, \ldots, r_{i,N-1}, i \rangle$ be a permutation of the sequence $\langle 1, \ldots, N \rangle$ such that $r_{i,N} = i$, and let $\bar{r}_{i,n} = \langle r_{i,1}, \ldots, r_{i,n} \rangle$, $n = 1, \ldots, N$ be the corresponding chain of growing sequences of states from space $\mathbb{X}$.

The following theorem is, in fact, a variant of Theorem 3.4. The above remarks let us, however, to simplify notations and reduce the proof.

Theorem 4.1. *Let conditions* **A**, **B**, $\mathbf{C}_1$, **D**, *and* $\mathbf{E}_1$ *hold for semi-Markov processes* $\eta_\varepsilon(t)$. *Then, for every* $i \in \mathbb{X}$, *the pivotal* $(M_{ii}^-[1], M_{ii}^+[1])$*-expansion for the expectation of return time* $E_{ii}(1,\varepsilon)$ *is given by the algorithm based on the sequential exclusion of states* $r_{i,1}, \ldots, r_{i,N-1}$ *from the phase space* $\mathbb{X}$ *of the processes* $\eta_\varepsilon(t)$. *This algorithm is described below, in the proof of the theorem. The above* $(M_{ii}^-[1], M_{ii}^+[1])$*-expansion is invariant with respect to any permutation* $\bar{r}_{i,N} = \langle r_{i,1}, \ldots, r_{i,N-1}, i \rangle$ *of sequence* $\langle 1, \ldots, N \rangle$.

Proof. Let us assume that $p_{\varepsilon,i} = 1$. Denote as ${}_{\bar{r}_{i,0}}\eta_\varepsilon(t) = \eta_\varepsilon(t)$, the initial semi-Markov process. Let us exclude state $r_{i,1}$ from the phase space of semi-Markov process ${}_{\bar{r}_{i,0}}\eta_\varepsilon(t)$ using the time-space screening procedure described in Section 3.2. Let ${}_{\bar{r}_{i,1}}\eta_\varepsilon(t)$ be the corresponding reduced semi-Markov process. The above procedure can be repeated. The state $r_{i,2}$ can be excluded from the phase space of the semi-Markov process ${}_{\bar{r}_{i,1}}\eta_\varepsilon(t)$. Let ${}_{\bar{r}_{i,2}}\eta_\varepsilon(t)$ be the corresponding reduced semi-Markov process. By continuing the above procedure for states $r_{i,3}, \ldots, r_{i,n}$, we construct the reduced semi-Markov process ${}_{\bar{r}_{i,n}}\eta_\varepsilon(t)$.

The process ${}_{\bar{r}_{i,n}}\eta_\varepsilon(t)$ has the phase space ${}_{\bar{r}_{i,n}}\mathbb{X} = \mathbb{X} \setminus \{r_{i,1}, r_{i,2}, \ldots, r_{i,n}\}$. The transition probabilities of the embedded Markov chain ${}_{\bar{r}_{i,n}}p_{i'j'}(\varepsilon), i', j' \in {}_{\bar{r}_{i,n}}\mathbb{X}$, and the expectations of transition times ${}_{\bar{r}_{i,n}}e_{i'j'}(1,\varepsilon), i', j' \in {}_{\bar{r}_{i,n}}\mathbb{X}$ are determined for the semi-Markov process ${}_{\bar{r}_{i,n}}\eta_\varepsilon(t)$ by the transition probabilities and the expectations of transition times for the process ${}_{\bar{r}_{i,n-1}}\eta_\varepsilon(t)$, respectively, via relations (3.31) and (3.39).

By Theorem 3.1, the expectation of hitting time $E_{i'j'}(1,\varepsilon)$ coincides for the semi-Markov processes ${}_{\bar{r}_{i,0}}\eta_\varepsilon(t), {}_{\bar{r}_{i,1}}\eta_\varepsilon(t), \ldots, {}_{\bar{r}_{i,n}}\eta_\varepsilon(t)$, for every $i', j' \in {}_{\bar{r}_{i,n}}\mathbb{X}$.

By Theorems 3.2 and 3.3 (the part related to the expectations), the semi-Markov processes ${}_{\bar{r}_{i,n}}\eta_\varepsilon(t)$ satisfy conditions **B**, $\mathbf{C}_1$ and, also, conditions **A**, **D**, and $\mathbf{E}_1$, with the same parameter ε_0 as for processes ${}_{\bar{r}_{i,n-1}}\eta_\varepsilon(t)$.

The transition sets ${}_{\bar{r}_{i,n}}\mathbb{Y}_{i'}$, $i' \in {}_{\bar{r}_{i,n}}\mathbb{X}$ are determined by the transition sets ${}_{\bar{r}_{i,n-1}}\mathbb{Y}_{i'}$, $i' \in {}_{\bar{r}_{i,n-1}}\mathbb{X}$, via relation (3.42) given in Lemma 3.2.

Therefore, the pivotal $({}_{\bar{r}_{i,n}}l^-_{i'j'}, {}_{\bar{r}_{i,n}}l^+_{i'j'})$-expansions,

$$ {}_{\bar{r}_{i,n}}p_{i'j'}(\varepsilon) = \sum_{{}_{\bar{r}_{i,n}}l^-_{i'j'}}^{{}_{\bar{r}_{i,n}}l^+_{i'j'}} {}_{\bar{r}_{i,n}}a_{i'j'}[l]\varepsilon^l + {}_{\bar{r}_{i,n}}o_{i'j'}(\varepsilon^{{}_{\bar{r}_{i,n}}l^+_{i'j'}}), \varepsilon \in (0, \varepsilon_0], \tag{4.1} $$

and the pivotal $({}_{\bar{r}_{i,n}}m^-_{i'j'}[1], {}_{\bar{r}_{i,n}}m^+_{i'j'}[1])$-expansions,

$$ {}_{\bar{r}_{i,n}}e_{i'j'}(1,\varepsilon) = \sum_{{}_{\bar{r}_{i,n}}m^-_{i'j'}[1]}^{{}_{\bar{r}_{i,n}}m^+_{i'j'}[1]} {}_{\bar{r}_{i,n}}b_{i'j'}[1,l]\varepsilon^l s + {}_{\bar{r}_{i,n}}\dot{o}_{1,i'j'}(\varepsilon^{{}_{\bar{r}_{i,n}}m^+_{i'j'}[1]}),\ \varepsilon \in (0, \varepsilon_0], \tag{4.2} $$

can be constructed for $j' \in {}_{\bar{r}_{i,n}}\mathbb{Y}_{i'}$, $i' \in {}_{\bar{r}_{i,n}}\mathbb{X}$ by applying the algorithms given in Theorems 3.2 and 3.3, respectively, to the $({}_{\bar{r}_{i,n-1}}l^-_{i'j'}, {}_{\bar{r}_{i,n-1}}l^+_{i'j'})$-expansions for transition probabilities ${}_{\bar{r}_{i,n-1}}p_{i'j'}(\varepsilon)$, $j' \in {}_{\bar{r}_{i,n-1}}\mathbb{Y}_{i'}$, $i' \in {}_{\bar{r}_{i,n-1}}\mathbb{X}$ and to the $({}_{\bar{r}_{i,n-1}}m^-_{i'j'}[1], {}_{\bar{r}_{i,n-1}}m^+_{i'j'}[1])$-expansions for expectations ${}_{\bar{r}_{i,n-1}}e_{i'j'}(1,\varepsilon)$, $j' \in {}_{\bar{r}_{i,n-1}}\mathbb{Y}_{i'}$, $i' \in {}_{\bar{r}_{i,n-1}}\mathbb{X}$.

The algorithm described above has a recurrent form and should be realized sequentially for the reduced semi-Markov processes ${}_{\bar{r}_{i,1}}\eta_\varepsilon(t), \ldots, {}_{\bar{r}_{i,n}}\eta_\varepsilon(t)$ starting from the initial semi-Markov process ${}_{\bar{r}_{i,0}}\eta_\varepsilon(t)$.

The asymptotic expansions for functions ${}_{\bar{r}_{i,n}}p_{i'j'}(\varepsilon)$ and ${}_{\bar{r}_{i,n}}e_{i'j'}(1,\varepsilon)$ are invariant with respect to any permutation $\bar{r}'_{i,n}$ of the sequence $\bar{r}_{i,n}$. This was proved in Theorem 3.4.

Let us take $n = N-1$. In this case, the semi-Markov process ${}_{\bar{r}_{i,N-1}}\eta^{(\varepsilon)}(t)$ has the phase space ${}_{\bar{r}_{i,N-1}}\mathbb{X} = \mathbb{X} \setminus \{r_{i,1}, r_{i,2}, \ldots, r_{i,N-1}\} = \{i\}$, which is a one-state set. The process ${}_{\bar{r}_{i,N-1}}\eta_\varepsilon(t)$ returns in state i after every jump. Its transition probability ${}_{\bar{r}_{i,N-1}}p_{ii}(\varepsilon) = 1$ and the expectation of return time,

$$ E_{ii}(1,\varepsilon) = {}_{\bar{r}_{i,N-1}}e_{ii}(1,\varepsilon). \tag{4.3} $$

Thus, the above recurrent algorithm of sequential phase space reduction makes it possible to write down the following pivotal Laurent asymptotic expansion:

$$ E_{ii}(1,\varepsilon) = \sum_{l=M^-_{ii}[1]}^{M^+_{ii}[1]} B_{ii}[1,l]\varepsilon^l + \ddot{o}_{1,ii}(\varepsilon^{M^+_{ii}[1]}),\ \varepsilon \in (0, \varepsilon_0], \tag{4.4} $$

where (a) $M^\pm_{ii}[1] = {}_{\bar{r}_{i,N-1}}m^\pm_{ii}[1]$; (b) $B_{ii}[1,l] = {}_{\bar{r}_{i,N-1}}b_{ii}[1,l]$, $l = M^-_{ii}[1], \ldots, M^+_{ii}[1]$; (c) $\ddot{o}_{1,ii}(\varepsilon^{M^+_{ii}[1]}) = {}_{\bar{r}_{i,N-1}}\dot{o}_{1,ii}(\varepsilon^{M^+_{ii}[1]})$.

By the above remarks, the asymptotic expansion given in relation (4.4) is invariant with respect to the choice of sequence $\bar{r}_{i,N-1} = \langle r_{i,1}, \ldots, r_{i,N-1}\rangle$. This legitimates notations (with omitted index ${}_{\bar{r}_{i,N-1}}$) used for parameters, coefficients, and remainder in the above asymptotic expansion.

The algorithm for construction of the Laurent asymptotic expansion for expectation $E_{ii}(1,\varepsilon)$, given in relation (4.4), can be repeated for every $i \in \mathbb{X}$. $\square$

Remark 4.1. Since matrices $\|_{\bar{r}_{i,n}}p_{i'j'}(\varepsilon)\|, \varepsilon \in (0,\varepsilon_0], n = 0,\dots,N-1$ are stochastic, the asymptotic expansions for transition probabilities $_{\bar{r}_{i,n}}p_{i'j'}(\varepsilon)$, $j' \in {}_{\bar{r}_{i,n}}\mathbb{Y}_{i'}, i' \in {}_{\bar{r}_{i,n}}\mathbb{X}$ satisfy condition **F**, for every $n = 0,\dots,N-1$.

Remark 4.2. Since the Laurent asymptotic expansions (4.4) are pivotal, the coefficients $B_{ii}[1, M_{ii}^-[1]] > 0$, for $i \in \mathbb{X}$.

4.1.2 Asymptotic Expansions with Remainders Given in the Standard Form for Stationary Distributions

The following theorem presents asymptotic expansions with remainders given in the standard form for stationary distributions of nonlinearly perturbed semi-Markov processes.

Theorem 4.2. *Let conditions* **A**, **B**, **C**$_1$, **D**, *and* **E**$_1$ *hold for semi-Markov processes* $\eta_\varepsilon(t)$. *Then, for every* $i \in \mathbb{X}$, *the pivotal* (n_i^-, n_i^+)*-expansion for the stationary probability* $\pi_i(\varepsilon)$ *is given by the algorithm based on the sequential exclusion of states* $r_{i,1},\dots,r_{i,N-1}$ *from the phase space* $\mathbb{X}$ *of the processes* $\eta_\varepsilon(t)$. *This algorithm is described below, in the proof of the theorem. The above* (n_i^-, n_i^+)*-expansion is invariant with respect to any permutation* $\bar{r}_{i,N} = \langle r_{i,1},\dots, r_{i,N-1}, i\rangle$ *of sequence* $\langle 1,\dots,N\rangle$. *Relations* **(1)**–**(6)**, *given in the proof, hold for these expansions.*

Proof. First, condition **E**$_1$ and proposition **(i)** (the multiple summation rule) of Lemma 2.3 make it possible to write down pivotal (m_i^-, m_i^+)-expansions for expectations $e_i(\varepsilon), i \in \mathbb{X}$. These expansions take the following form, for $i \in \mathbb{X}$:

$$e_i(\varepsilon) = \sum_{j \in \mathbb{Y}_i} e_{ij}(1,\varepsilon) = \sum_{l=m_i^-}^{m_i^+} b_i[l]\varepsilon^l + \dot{o}_i(\varepsilon^{m_i^+}),\ \varepsilon \in (0,\varepsilon_0], \tag{4.5}$$

where (a) $m_i^\pm = \min_{j \in \mathbb{Y}_i} m_{ij}^\pm[1]$; (b) $b_i[m_i^- + l] = \sum_{j \in Y_i} b_{ij}[1, m_i^- + l]$, $l = 0,\dots,m_i^+ - m_i^-$, where $b_{ij}[1, m_i^- + l] = 0$, for $0 \le l < m_{ij}^- - m_i^-, j \in Y_i$; (c) $\dot{o}_i(\varepsilon^{m_i^+})$ is given by formula **(c)** from proposition **(i)** (the multiple summation rule) of Lemma 2.3, which should be applied to the corresponding Laurent asymptotic expansions given in condition **E**$_1$.

Second, conditions **A**, **B**, **C**$_1$, **D**, and **E**$_1$, the asymptotic expansions given in relations (4.4) and (4.5), and proposition **(v)** (the division rule) of Lemma 2.2 make it possible to write down (n_i^-, n_i^+)-expansions for the stationary probabilities $\pi_i(\varepsilon) = \frac{e_i(\varepsilon)}{E_{ii}(\varepsilon)}, i \in \mathbb{X}$. These expansions take the following form, for $i \in \mathbb{X}$:

$$\pi_i(\varepsilon) = \sum_{l=n_i^-}^{n_i^+} c_i[l]\varepsilon^l + o_i(\varepsilon^{n_i^+}),\ \varepsilon \in (0, \varepsilon_0], \tag{4.6}$$

where: (a) $n_i^- = m_i^- - M_{ii}^-[1]$, $n_i^+ = (m_i^+ - M_{ii}^-[1]) \wedge (M_{ii}^+[1] - 2M_{ii}^-[1] + m_i^-)$; (b) $c_i[n_i^- + l] = B_{ii}[1, M_{ii}^-[1]]^{-1}(b_i[m_i^- + l] - \sum_{1 \le l' \le l} B_{ii}[1, M_{ii}^-[1] + l]\, c_i[n_i^- + l - l'])$, $l = 0, \dots, n_i^+ - n_i^-$; (c) $o_i(\varepsilon^{n_i^+})$ is given by formula **(f)** from proposition **(v)** (the division rule) of Lemma 2.2, which should be applied to the asymptotic expansions given in relations (4.4) and (4.5).

Since the asymptotic expansions given in relations (4.4) and (4.5) are pivotal, the expansions given in relation (4.6) are also pivotal, i.e., $c_i[n_i^-] = b_i[m_i^-]/B_{ii}[1, M_{ii}^-[1]] \neq 0,\ i \in \mathbb{X}$. Moreover, since $\pi_i(\varepsilon) > 0, i \in \mathbb{X}$, $\varepsilon \in (0, \varepsilon_0]$, the following relation takes place: **(1)** $c_i[n_i^-] > 0,\ i \in \mathbb{X}$.

By the definition, $e_i(\varepsilon) \le E_{ii}(1, \varepsilon),\ i \in \mathbb{X},\ \varepsilon \in (0, \varepsilon_0]$. This implies that parameters $M_{ii}^-[1] \le m_i^-, i \in \mathbb{X}$ and, thus, **(2)** $n_i^- \ge 0,\ i \in \mathbb{X}$.

Since, $\sum_{i \in \mathbb{X}} \pi_i(\varepsilon) = 1$, $\varepsilon \in (0, \varepsilon_0]$, parameters $n_i^\pm$, $i \in \mathbb{X}$ and coefficients $c_i[l]$, $l = n_i^-, \dots, n_i^+, i \in \mathbb{X}$ satisfy relations, **(3)** $n^- = \min_{i \in \mathbb{X}} n_i^- = 0$, and, **(4)** $\sum_{i \in \mathbb{X}} c_i[l] = \mathrm{I}(l = 0)$, $0 \le l \le n^+ = \min_{i \in \mathbb{X}} n_i^+$. Moreover, the remainders of asymptotic expansions given in (4.6) satisfy identity, **(5)** $\sum_{i \in \mathbb{X}}(\sum_{n^+ < l \le n_i^+} c_i[l]\varepsilon^l + o_i(\varepsilon^{n_i^+})) = 0$, $\varepsilon \in (0, \varepsilon_0]$.

By the above remarks, **(6)** there exists $\lim_{\varepsilon \to 0} \pi_i(\varepsilon) = \pi_i(0)$, which equals to $c_i[0] > 0$ if $i \in \mathbb{X}_0$, or 0 if $i \notin \mathbb{X}_0$, where $\mathbb{X}_0 = \{i \in \mathbb{X} : n_i^- = 0\}$.

As follows from Theorem 4.1, the asymptotic expansion (4.4) for expectation $E_{ii}(1, \varepsilon)$ and, thus, the asymptotic expansion (4.6) for stationary probability $\pi_i(\varepsilon)$ is, for every $i \in \mathbb{X}$, invariant with respect to any permutation $\bar{r}_{i,N} = \langle r_{i,1}, \dots, r_{i,N-1}, i \rangle$ of sequence $\langle 1, \dots, N \rangle$. □

4.2 Asymptotic Expansions with Explicit Upper Bounds for Remainders for Stationary Distributions of Perturbed Semi-Markov Processes

In this section, we describe an algorithm for construction of asymptotic expansions with explicit upper bounds for remainders for stationary distributions of nonlinearly perturbed semi-Markov processes.

4.2.1 Asymptotic Expansions with Explicit Upper Bounds for Remainders for Expectations of Return Times

As in Subsection 4.1.1, let $\bar{r}_{i,N} = \langle r_{i,1},\dots,r_{i,N}\rangle = \langle r_{i,1},\dots,r_{i,N-1},i\rangle$ be a permutation of the sequence $\langle 1,\dots,N\rangle$ such that $r_{i,N} = i$, and let $\bar{r}_{i,n} = \langle r_{i,1},\dots,r_{i,n}\rangle, n = 1,\dots,N$ be the corresponding chain of growing sequences of states from space $\mathbb{X}$.

Theorem 4.3. *Let conditions* **A**, **B**, $\mathbf{C}_1$, $\mathbf{D}'$, *and* $\mathbf{E}'_1$ *hold for the semi-Markov processes* $\eta_\varepsilon(t)$. *Then, for every* $i \in \mathbb{X}$, *the pivotal* $(M^-_{ii}[1], M^+_{ii}[1])$*-expansion for the expectation of return time* $E_{ii}(1,\varepsilon)$, *given in Theorem 4.1 and obtained as the result of sequential exclusion of states* $r_{i,1},\dots,r_{i,N-1}$ *from the phase space* $\mathbb{X}$ *of the processes* $\eta_\varepsilon(t)$, *is a* $(M^-_{ii}[1], M^+_{ii}[1], {}_{\bar{r}_{i,N-1}}\dot{\delta}_{ii}[1], {}_{\bar{r}_{i,N-1}}\dot{G}_{ii}[1], {}_{\bar{r}_{i,N-1}}\dot{\varepsilon}_{ii}[1])$*-expansion. Parameters* ${}_{\bar{r}_{i,N-1}}\dot{\delta}_{ii}[1]$, ${}_{\bar{r}_{i,N-1}}\dot{G}_{ii}[1]$, *and* ${}_{\bar{r}_{i,N-1}}\dot{\varepsilon}_{ii}[1]$ *can be computed using the algorithm described below, in the proof of the theorem.*

Proof. Let us assume that $p_{\varepsilon,i} = 1$. Denote as ${}_{\bar{r}_{i,0}}\eta_\varepsilon(t) = \eta_\varepsilon(t)$ the initial semi-Markov process. Let us exclude state $r_{i,1}$ from the phase space of semi-Markov process ${}_{\bar{r}_{i,0}}\eta_\varepsilon(t)$ using the time-space screening procedure described in Section 3.2. Let ${}_{\bar{r}_{i,1}}\eta_\varepsilon(t)$ be the corresponding reduced semi-Markov process. The above procedure can be repeated. The state $r_{i,2}$ can be excluded from the phase space of the semi-Markov process ${}_{\bar{r}_{i,1}}\eta_\varepsilon(t)$. Let ${}_{\bar{r}_{i,2}}\eta_\varepsilon(t)$ be the corresponding reduced semi-Markov process. By continuing the above procedure for states $r_{i,3},\dots,r_{i,n}$, we construct the reduced semi-Markov process ${}_{\bar{r}_{i,n}}\eta_\varepsilon(t)$.

The process ${}_{\bar{r}_{i,n}}\eta_\varepsilon(t)$ has the phase space ${}_{\bar{r}_{i,n}}\mathbb{X} = \mathbb{X} \setminus \{r_{i,1}, r_{i,2},\dots,r_{i,n}\}$. The transition probabilities ${}_{\bar{r}_{i,n}}p_{i'j'}(\varepsilon)$, $i',j' \in {}_{\bar{r}_{i,n}}\mathbb{X}$ and the expectations of sojourn times ${}_{\bar{r}_{i,n}}e_{i'j'}(1,\varepsilon), i',j' \in {}_{\bar{r}_{i,n}}\mathbb{X}$ are determined for the process ${}_{\bar{r}_{i,n}}\eta_\varepsilon(t)$ by the transition probabilities and the expectations of sojourn times for the process ${}_{\bar{r}_{i,n-1}}\eta_\varepsilon(t)$, via relations (3.31) and (3.39).

By Theorem 3.1, the expectation of hitting time $E_{i'j'}(1,\varepsilon)$ coincides for the semi-Markov processes ${}_{\bar{r}_{i,0}}\eta_\varepsilon(t)$, ${}_{\bar{r}_{i,1}}\eta_\varepsilon(t),\dots$, ${}_{\bar{r}_{i,n}}\eta_\varepsilon(t)$, for every $i',j' \in {}_{\bar{r}_{i,n}}\mathbb{X}$.

By Theorems 3.2, 3.3, 3.5, and 3.6, the semi-Markov processes ${}_{\bar{r}_{i,n}}\eta_\varepsilon(t)$ satisfy conditions **A**, **B**, $\mathbf{C}_1$, $\mathbf{D}'$, and $\mathbf{E}'_1$.

The transition sets ${}_{\bar{r}_{i,n}}\mathbb{Y}_{i'}, i' \in {}_{\bar{r}_{i,n}}\mathbb{X}$, for the process ${}_{\bar{r}_{i,n}}\eta_\varepsilon(t)$, are determined by the transition sets ${}_{\bar{r}_{i,n-1}}\mathbb{Y}_{i'}, i' \in {}_{\bar{r}_{i,n-1}}\mathbb{X}$, for the process ${}_{\bar{r}_{i,n-1}}\eta_\varepsilon(t)$, via relation (3.42) given in Lemma 3.2.

For every $j' \in {}_{\bar{r}_{i,n}}\mathbb{Y}_{i'}, i' \in {}_{\bar{r}_{i,n}}\mathbb{X}$, the pivotal $({}_{\bar{r}_{i,n}}l^-_{i'j'}, {}_{\bar{r}_{i,n}}l^+_{i'j'})$-expansion for transition probability ${}_{\bar{r}_{i,n}}p_{i'j'}(\varepsilon)$, given in Theorem 3.2, is, by Theorem 3.5, a $({}_{\bar{r}_{i,n}}l^-_{i'j'}, {}_{\bar{r}_{i,n}}l^+_{i'j'}, {}_{\bar{r}_{i,n}}\delta_{i'j'}, {}_{\bar{r}_{i,n}}G_{i'j'}, {}_{\bar{r}_{i,n}}\varepsilon_{i'j'})$-expansion, with parameters ${}_{\bar{r}_{i,n}}\delta_{i'j'}$, ${}_{\bar{r}_{i,n}}G_{i'j'}$, and ${}_{\bar{r}_{i,n}}\varepsilon_{i'j'}$ given in this theorem. Analogously, for every $j' \in {}_{\bar{r}_{i,n}}\mathbb{Y}_{i'}, i' \in {}_{\bar{r}_{i,n}}\mathbb{X}$, the pivotal $({}_{\bar{r}_{i,n}}m^-_{i'j'}[1], {}_{\bar{r}_{i,n}}m^+_{i'j'}[1])$-expansion for expectation ${}_{\bar{r}_{i,n}}e_{i'j'}(1,\varepsilon)$, given in Theorem 3.3, is, by Theorem 3.6, a $({}_{\bar{r}_{i,n}}m^-_{i'j'}[1], {}_{\bar{r}_{i,n}}m^+_{i'j'}[1], {}_{\bar{r}_{i,n}}\dot{\delta}_{i'j'}[1], {}_{\bar{r}_{i,n}}\dot{G}_{i'j'}[1], {}_{\bar{r}_{i,n}}\dot{\varepsilon}_{i'j'}[1])$-expansion, with parameters ${}_{\bar{r}_{i,n}}\dot{\delta}_{i'j'}[1]$, ${}_{\bar{r}_{i,n}}\dot{G}_{i'j'}[1]$, and ${}_{\bar{r}_{i,n}}\dot{\varepsilon}_{i'j'}[1]$ given in this theorem. Also, by Theorem 3.6, the inequalities ${}_{\bar{r}_{i,n}}\dot{\delta}_{i'j'}[1] \geq \delta^*[1], j' \in {}_{\bar{r}_{i,n}}\mathbb{Y}_{i'}, i' \in {}_{\bar{r}_{i,n}}\mathbb{X}$ hold.

The algorithm described above has a recurrent form and should be realized sequentially for the reduced semi-Markov processes ${}_{\bar{r}_{i,1}}\eta_\varepsilon(t), \ldots, {}_{\bar{r}_{i,n}}\eta_\varepsilon(t)$ starting from the initial semi-Markov process ${}_{\bar{r}_{i,0}}\eta_\varepsilon(t)$.

Let us take $n = N-1$. The semi-Markov process ${}_{\bar{r}_{i,N-1}}\eta_\varepsilon(t)$ has the phase space ${}_{\bar{r}_{i,N-1}}\mathbb{X} = \mathbb{X} \setminus \{r_{i,1}, r_{i,2}, \ldots, r_{i,N-1}\} = \{i\}$, which is a one-state set. The process ${}_{\bar{r}_{i,N-1}}\eta_\varepsilon(t)$ returns to state i after every jump. Its transition probability ${}_{\bar{r}_{i,N-1}}p_{ii}(\varepsilon) = 1$, and the expectation of return time, $E_{ii}(1,\varepsilon) = {}_{\bar{r}_{i,N-1}}e_{ii}(1,\varepsilon)$. This equality and Theorem 3.4 yield, for every $i \in \mathbb{X}$, the pivotal $(M_{ii}^-[1], M_{ii}^+[1])$-expansion for expectation $E_{ii}(1,\varepsilon)$, which is invariant with respect to any permutation $\bar{r}'_{i,N-1} = \langle r'_{i,1}, \ldots, r'_{i,N-1}\rangle$ of the sequence $\bar{r}_{i,N-1} = \langle r_{i,1}, \ldots, r_{i,N-1}\rangle$. This invariance also implies that parameters ${}_{\bar{r}_{i,N-1}}m_{ii}^\pm[1] = M_{ii}^\pm[1]$ do not depend on the choice of sequence $\bar{r}_{i,N-1}$. By Theorem 3.6, the above $(M_{ii}^-[1], M_{ii}^+[1])$-expansion for $E_{ii}(1,\varepsilon) = {}_{\bar{r}_{i,N-1}}e_{ii}(1,\varepsilon)$ is a $(M_{ii}^-[1], M_{ii}^+[1], {}_{\bar{r}_{i,N-1}}\dot{\delta}_{ii}[1], {}_{\bar{r}_{i,N-1}}\dot{G}_{ii}[1], {}_{\bar{r}_{i,N-1}}\dot{\varepsilon}_{ii}[1])$-expansion.

The above algorithm can be realized for any sequence $\bar{r}_{i,N-1} = \langle r_{i,1}, \ldots, r_{i,N-1}\rangle$, but the invariance of explicit upper bounds for remainders, with respect to permutations $\bar{r}_{i,N} = \langle r_{i,1}, \ldots, r_{i,N-1}, i\rangle$ of sequence $\langle 1, \ldots, N\rangle$, cannot be guaranteed.

The algorithm described above can be repeated, for every $i \in \mathbb{X}$. □

Remark 4.3. By Theorem 3.7, inequality ${}_{\bar{r}_{i,N-1}}\dot{\delta}_{ii}[1] \geq \delta^*[1]$ holds, for any sequence $\bar{r}_{i,N-1} = \langle r_{i,1}, \ldots, r_{i,N-1}\rangle$. This makes it possible to rewrite function $E_{ii}(1,\varepsilon)$ as the pivotal $(M_{ii}^-[1], M_{ii}^+[1], \delta^*[1], {}_{\bar{r}_{i,N-1}}G_{ii}^*[1], {}_{\bar{r}_{i,N-1}}\dot{\varepsilon}_{ii}[1])$-expansion, with parameter ${}_{\bar{r}_{i,N-1}}G_{ii}^*[1] = {}_{\bar{r}_{i,N-1}}\dot{G}_{ii}[1] \cdot ({}_{\bar{r}_{i,N-1}}\dot{\varepsilon}_{ii}[1])^{({}_{\bar{r}_{i,N-1}}\dot{\delta}_{ii}[1] - \delta^*[1])}$.

4.2.2 Asymptotic Expansions with Explicit Upper Bounds for Remainders for Stationary Distributions

Let us recall the pivotal (n_i^-, n_i^+)-expansion for stationary probability $\pi_i(\varepsilon)$ of nonlinearly perturbed semi-Markov process $\eta^{(\varepsilon)}(t)$ given, under conditions **A**, **B**, $\mathbf{C}_1$, **D**, and $\mathbf{E}_1$, in Theorem 4.2. This asymptotic expansion has the following form, for $i \in \mathbb{X}$:

$$\pi_i(\varepsilon) = \sum_{l=n_i^-}^{n_i^+} c_i[l]\varepsilon^l + o_i(\varepsilon^{n_i^+}),\ \varepsilon \in (0, \varepsilon_0]. \tag{4.7}$$

According to Theorem 4.2, the above asymptotic expansion is invariant with respect to the choice of sequence states $\bar{r}_{i,N-1} = (r_{i,1}, \ldots, r_{i,N-1})$ used in the corresponding algorithm, for every $i \in \mathbb{X}$.

The following theorem presents asymptotic expansions with explicit upper bounds for remainders, for stationary distributions of nonlinearly perturbed semi-Markov processes.

Theorem 4.4. *Let conditions* **A**, **B**, $\mathbf{C}_1$, $\mathbf{D}'$, *and* $\mathbf{E}'_1$ *hold for the semi-Markov processes* $\eta_\varepsilon(t)$. *Then, for every* $i \in \mathbb{X}$, *the pivotal* (n_i^-, n_i^+)*-expansion* (4.7) *for the stationary probability* $\pi_i(\varepsilon)$, *given in Theorem 4.2 and obtained as the result of sequential exclusion of states* $r_{i,1}, \ldots, r_{i,N-1}$ *from the phase space* $\mathbb{X}$ *of the processes* $\eta_\varepsilon(t)$, *is a* $(n_i^-, n_i^+, {}_{\bar{r}_{i,N-1}}\delta_i, {}_{\bar{r}_{i,N-1}}G_i, {}_{\bar{r}_{i,N-1}}\varepsilon_i)$*-expansion. Parameters* ${}_{\bar{r}_{i,N-1}}\delta_i$, ${}_{\bar{r}_{i,N-1}}G_i$, *and* ${}_{\bar{r}_{i,N-1}}\varepsilon_i$ *can be computed using the algorithm described below, in the proof of the theorem.*

Proof. Let us choose an arbitrary state $i \in \mathbb{X}$.

First, proposition **(i)** (the multiple summation rule) of Lemmas 2.3 and 2.7 should be applied to the pivotal (m_i^-, m_i^+)-expansion for the expectation $e_i(\varepsilon) = \sum_{j \in \mathbb{Y}_i} e_{ij}(\varepsilon)$ given by relation (4.5), in the proof of Theorem 4.2. This yields a $(m_i^-, m_i^+, \dot{\delta}_i, \dot{G}_i, \dot{\varepsilon}_i)$-expansion for expectation $e_i(\varepsilon)$, with the corresponding parameters $\dot{\delta}_i, \dot{G}_i$, and $\dot{\varepsilon}_i$.

Second, the propositions **(v)** (the division rule) of Lemmas 2.2 and 2.6 should be applied to the quotient $\pi_i(\varepsilon) = \frac{e_i(\varepsilon)}{E_{ii}(1,\varepsilon)}$. The $(m_i^-, m_i^+, \dot{\delta}_i, \dot{G}_i, \dot{\varepsilon}_i)$-expansion for expectation $e_i(\varepsilon)$ and the $(M_{ii}^-[1], M_{ii}^+[1], {}_{\bar{r}_{i,N-1}}\dot{\delta}_{ii}[1], {}_{\bar{r}_{i,N-1}}\dot{G}_{ii}[1], {}_{\bar{r}_{i,N-1}}\dot{\varepsilon}_{ii})[1]$-expansion for the expectation of hitting time $E_{ii}(1,\varepsilon)$, given in Theorems 4.1 and 4.3, should be used. This yields the corresponding pivotal (n_i^-, n_i^+)-expansion for the stationary probability $\pi_i(\varepsilon)$, given in Theorem 4.2, and proves that this expansion is a $(n_i^-, n_i^+, {}_{\bar{r}_{i,N-1}}\delta_i, {}_{\bar{r}_{i,N-1}}G_i, {}_{\bar{r}_{i,N-1}}\varepsilon_i)$-expansion, with parameters computed in the process of realization of the above algorithm. □

The explicit upper bounds for remainders in the asymptotic expansions given in Theorem 4.4 have a clear and informative power-type form. An useful property of these upper bounds is that they are asymptotically uniform with respect to the perturbation parameter. The recurrent algorithm for finding these upper bounds is computationally effective.

Unfortunately, the summation and multiplication operational rules for computing power-type upper bounds for remainders possess commutative but do not possess associative and distributive properties. This causes dependence of the resulting upper bounds for remainders in the asymptotic expansions for stationary probabilities $\pi_i(\varepsilon), i \in \mathbb{X}$ on a choice of the corresponding sequences of states $\bar{r}_{i,N-1} = \langle r_{i,1}, \ldots, r_{i,N-1} \rangle, i \in \mathbb{X}$ used in the above algorithm. This rises two open questions, the first one, about possible alternative forms for remainders possessing the desirable algebraic properties mentioned above, and, the second one, about an optimal choice of sequences of states $\bar{r}_{i,N-1}, i \in \mathbb{X}$.

Remark 4.4. By proposition **(iii)** of Lemma 2.7, inequality ${}_{\bar{r}_{i,N-1}}\delta_i \geq \delta^*[1]$ holds, for every sequence $\bar{r}_{i,N-1}$, This makes it possible to rewrite function $\pi_i(\varepsilon)$ as the pivotal $(n_i^-, n_i^+, \delta^*[1], {}_{\bar{r}_{i,N-1}}G_i^*, {}_{\bar{r}_{i,N-1}}\varepsilon_i)$-expansion, with parameter ${}_{\bar{r}_{i,N-1}}G_i^* = {}_{\bar{r}_{i,N-1}}G_i \cdot ({}_{\bar{r}_{i,N-1}}\varepsilon_i)^{({}_{\bar{r}_{i,N-1}}\delta_i - \delta^*[1])}$.

4.3 Asymptotic Expansions for Stationary Functionals for Perturbed Semi-Markov Processes

In this section, we construct asymptotic expansions for some stationary functionals connected with stationary distributions of nonlinearly perturbed semi-Markov processes.

4.3.1 Asymptotic Expansions for Conditional Quasi-Stationary Distributions

Let us introduce the so-called conditional quasi-stationary distributions, which can be defined, for any nonempty subset $\mathbb{U} \subseteq \mathbb{X}$ by the following asymptotic relation, which takes place under conditions **A**, **B**, and $\mathbf{C}_1$, for every $\varepsilon \in (0, \varepsilon_0]$,

$$\begin{aligned}\bar{\mu}_{\varepsilon,\mathbb{U},i}(t) = \frac{\bar{\mu}_{\varepsilon,i}(t)}{\sum_{j\in\mathbb{U}}\bar{\mu}_{\varepsilon,j}(t)} &= \frac{\int_0^t I(\eta_\varepsilon(s) = i)ds}{\int_0^t I(\eta_\varepsilon(s) \in \mathbb{U})ds} \\ &\xrightarrow{a.s.} \pi_{\mathbb{U},i}(\varepsilon) = \frac{\pi_i(\varepsilon)}{\sum_{j\in\mathbb{U}}\pi_j(\varepsilon)} \ \text{as } t \to \infty, \text{ for } i \in \mathbb{U}. \end{aligned} \tag{4.8}$$

The conditional quasi-stationary probability counts the stationary average share of time spent by the semi-Markov process $\eta_\varepsilon(t)$ in state $i \in \mathbb{U}$ out of the stationary average time spent by this process in the subset of states $\mathbb{U}$.

The ergodic relation (4.8), which is the direct corollary of relation (3.17), holds for any initial distribution $\bar{p}^{(\varepsilon)}$ and the conditional quasi-stationary stationary distribution $\langle \pi_{\mathbb{U},i}(\varepsilon), i \in \mathbb{U}\rangle$ does not depend on the initial distribution. Also quasi-stationary probabilities $\pi_{\mathbb{U},i}(\varepsilon) > 0, i \in \mathbb{U}$, for $\varepsilon \in (0, \varepsilon_0]$.

The following theorem is a direct corollary of Theorems 4.2, which can be obtained by application of operational rules for Laurent asymptotic expansions given in Lemmas 2.1–2.4 to the quotient formula (4.8) for conditional quasi-stationary probabilities.

Theorem 4.5. *Let conditions* **A**, **B**, $\mathbf{C}_1$, **D**, *and* $\mathbf{E}_1$ *hold for semi-Markov processes* $\eta_\varepsilon(t)$. *Then, for every nonempty* $\mathbb{U} \subseteq \mathbb{X}$ *and* $i \in \mathbb{U}$, *the pivotal* $(n^-_{\mathbb{U},i}, n^+_{\mathbb{U},i})$*-expansion for the stationary probability* $\pi_{\mathbb{U},i}(\varepsilon)$ *is given by the algorithm described below, in the proof of the theorem. Relations* **(1)–(6)**, *given in the proof, hold for these expansions.*

Proof. Let us choose an arbitrary nonempty $\mathbb{U} \subseteq \mathbb{X}$ and state $i \in \mathbb{U}$.

First, Theorem 4.2 let us to apply proposition **(i)** (the multiple summation rule) of Lemmas 2.3 the sum $\Pi_{\mathbb{U}}(\varepsilon) = \sum_{j\in\mathbb{U}}\pi_j(\varepsilon)$ and to get the following asymptotic expansion:

$$\Pi_{\mathbb{U}}(\varepsilon) = \sum_{j\in\mathbb{U}} \pi_j(\varepsilon) = \sum_{l=n_{\mathbb{U}}^-}^{n_{\mathbb{U}}^+} C_{\mathbb{U}}[l]\varepsilon^l + \dot{o}_{\mathbb{U}}(\varepsilon^{n_{\mathbb{U}}^+}),\ \varepsilon \in (0,\varepsilon_0], \tag{4.9}$$

where (a) $n_{\mathbb{U}}^{\pm} = \min_{j\in\mathbb{U}} n_i^{\pm}$; (b) $C_{\mathbb{U}}[n_{\mathbb{U}}^- + l] = \sum_{j\in\mathbb{U}} c_i[n_{\mathbb{U}}^- + l],\ l = 0,\dots,n_{\mathbb{U}}^+ - n_{\mathbb{U}}^-$, where $c_i[n_{\mathbb{U}}^- + l] = 0$, for $0 \le l < n_i^- - n_{\mathbb{U}}^-, j \in \mathbb{U}$; (c) $\dot{o}_{\mathbb{U}}(\varepsilon^{n_{\mathbb{U}}^+})$ is given by formula **(c)** from proposition **(i)** (the multiple summation rule) of Lemma 2.3, which should be applied to the corresponding Laurent asymptotic expansions for stationary probabilities $\pi_i(\varepsilon), i \in \mathbb{U}$ given in relation (4.6).

Second, conditions **A**, **B**, $\mathbf{C}_1$, **D**, and $\mathbf{E}_1$, the asymptotic expansions given in relations (4.6) and (4.9), and proposition **(v)** (the division rule) of Lemma 2.2 make it possible to write down $(n_{\mathbb{U},i}^-, n_{\mathbb{U},i}^+)$-expansions for the quasi-stationary probabilities $\pi_{\mathbb{U},i}(\varepsilon) = \frac{\pi_i(\varepsilon)}{\Pi_{\mathbb{U}}(\varepsilon)}, i \in \mathbb{U}$. These expansions take the following form, for $i \in \mathbb{U}$:

$$\pi_{\mathbb{U},i}(\varepsilon) = \sum_{l=n_{\mathbb{U},i}^-}^{n_{\mathbb{U},i}^+} c_{\mathbb{U},i}[l]\varepsilon^l + o_{\mathbb{U},i}(\varepsilon^{n_{\mathbb{U},i}^+}),\ \varepsilon \in (0,\varepsilon_0], \tag{4.10}$$

where: (a) $n_{\mathbb{U},i}^- = n_i^- - n_{\mathbb{U}}^-$, $n_{\mathbb{U},i}^+ = (n_i^+ - n_{\mathbb{U}}^-) \wedge (n_{\mathbb{U}}^+ - 2n_{\mathbb{U}}^- + n_i^-)$; (b) $c_{\mathbb{U},i}[n_{\mathbb{U},i}^- + l] = C_{\mathbb{U}}[n_{\mathbb{U}}^-]^{-1}(c_i[n_i^- + l] - \sum_{1\le l'\le l} C_{\mathbb{U}}[n_{\mathbb{U}}^- + l]\, c_{\mathbb{U},i}[n_{\mathbb{U},i}^- + l - l']), l = 0,\dots,n_i^+ - n_i^-$; (c) $o_i(\varepsilon^{n_i^+})$ is given by formula **(f)** from proposition **(v)** (the division rule) of Lemma 2.2, which should be applied to the asymptotic expansions given in relations (4.6) and (4.9).

Since the asymptotic expansions given in relations (4.6) and (4.9) are pivotal, the expansions given in relation (4.10) are also pivotal, i.e., $c_{\mathbb{U},i}[n_{\mathbb{U},i}^-] = c_i[n_i^-]/C_{\mathbb{U}}[n_{\mathbb{U}}^-] \neq 0,\ i \in \mathbb{X}$. Moreover, since $\pi_{\mathbb{U},i}(\varepsilon) > 0, i \in \mathbb{X}, \varepsilon \in (0,\varepsilon_0]$, the following relation takes place: **(1)** $c_{\mathbb{U},i}[n_{\mathbb{U},i}^-] > 0,\ i \in \mathbb{X}$.

By the definition, $\pi_i(\varepsilon) \le \Pi_{\mathbb{U}}(\varepsilon),\ i \in \mathbb{U},\ \varepsilon \in (0,\varepsilon_0]$. This implies that parameters $n_{\mathbb{U}}^- \le n_i^-, i \in \mathbb{U}$ and, thus, **(2)** $n_{\mathbb{U},i}^- \ge 0,\ i \in \mathbb{X}$.

Since, $\sum_{i\in\mathbb{U}} \pi_{\mathbb{U},i}(\varepsilon) = 1,\ \varepsilon \in (0,\varepsilon_0]$, parameters $n_{\mathbb{U},i}^{\pm},\ i \in \mathbb{U}$ and coefficients $c_{\mathbb{U},i}[l]$, $l = n_{\mathbb{U},i}^-,\dots,n_{\mathbb{U},i}^+, i \in \mathbb{U}$ satisfy relations, **(3)** $n_{\mathbb{U}}^- = \min_{i\in\mathbb{U}} n_{\mathbb{U},i}^- = 0$, and, **(4)** $\sum_{i\in\mathbb{U}} c_{\mathbb{U},i}[l] = \mathrm{I}(l = 0),\ 0 \le l \le n_{\mathbb{U}}^+ = \min_{i\in\mathbb{X}} n_{\mathbb{U},i}^+$. Moreover, the remainders of asymptotic expansions given in (4.6) satisfy identity, **(5)** $\sum_{i\in\mathbb{U}}(\sum_{n_{\mathbb{U}}^+ < l \le n_i^+} c_{\mathbb{U},i}[l]\varepsilon^l + o_{\mathbb{U},i}(\varepsilon^{n_{\mathbb{U},i}^+})) = 0,\ \varepsilon \in (0,\varepsilon_0]$.

By the above remarks, **(6)** there exists $\lim_{\varepsilon\to 0} \pi_{\mathbb{U},i}(\varepsilon) = \pi_{\mathbb{U},i}(0)$, which equals to $c_{\mathbb{U},i}[0] > 0$ if $i \in \mathbb{U}_0$, or 0 if $i \notin \mathbb{U}_0$, where $\mathbb{U}_0 = \{i \in \mathbb{U} : n_{\mathbb{U},i}^- = 0\}$.

The following theorem is a direct corollary of Theorems 4.4, which can be obtained by application of operational rules for Laurent asymptotic expansions given in Lemmas 2.1–2.8 to the quotient formula (4.8) for conditional quasi-stationary probabilities.

Theorem 4.6. *Let conditions* **A**, **B**, $\mathbf{C}_1$, $\mathbf{D}'$, *and* $\mathbf{E}'_1$ *hold for the semi-Markov processes* $\eta_\varepsilon(t)$. *Then, for every nonempty* $\mathbb{U} \subseteq \mathbb{X}$ *and* $i \in \mathbb{U}$, *the pivotal* $(n^-_{\mathbb{U},i}, n^+_{\mathbb{U},i})$-*expansion* (4.10) *for the conditional quasi-stationary probability* $\pi_{\mathbb{U},i}(\varepsilon)$, *given in Theorem 4.5, is a* $(n^-_{\mathbb{U},i}, n^+_{\mathbb{U},i}, {}_{\bar{r}_{i,N-1}}\delta_{\mathbb{U},i}, {}_{\bar{r}_{i,N-1}}G_{\mathbb{U},i}, {}_{\bar{r}_{i,N-1}}\varepsilon_{\mathbb{U},i})$-*expansion. Parameters* ${}_{\bar{r}_{i,N-1}}\delta_{\mathbb{U},i}$, ${}_{\bar{r}_{i,N-1}}G_{\mathbb{U},i}$, *and* ${}_{\bar{r}_{i,N-1}}\varepsilon_{\mathbb{U},i}$ *can be computed using the algorithm described below, in the proof.*

Proof. Let us choose an arbitrary nonempty $\mathbb{U} \subseteq \mathbb{X}$ and state $i \in \mathbb{X}$.

First, proposition **(i)** (the multiple summation rule) of Lemmas 2.3 and 2.7 should be applied to the pivotal $(n^-_{\mathbb{U},i}, n^+_{\mathbb{U},i})$-expansion for the sum $\Pi_{\mathbb{U}}(\varepsilon) = \sum_{j \in \mathbb{U}} \pi_j(\varepsilon)$ given by relation (4.9), in the proof of Theorem 4.5. This yields a $(n^-_{\mathbb{U}}, n^+_{\mathbb{U}}, {}_{\bar{r}_{i,N-1}}\delta_{\mathbb{U}}, {}_{\bar{r}_{i,N-1}}G_{\mathbb{U}}, {}_{\bar{r}_{i,N-1}}\varepsilon_{\mathbb{U}})$-expansion for function $\Pi_{\mathbb{U}}(\varepsilon)$, with the corresponding parameters ${}_{\bar{r}_{i,N-1}}\delta_{\mathbb{U}}$, ${}_{\bar{r}_{i,N-1}}G_{\mathbb{U}}$, and ${}_{\bar{r}_{i,N-1}}\varepsilon_{\mathbb{U}}$.

Second, the propositions **(v)** (the division rule) of Lemmas 2.2 and 2.6 should be applied to the quotient $\pi_{\mathbb{U},i}(\varepsilon) = \frac{\pi_i(\varepsilon)}{\Pi_{\mathbb{U}}(\varepsilon)}$. The $(n^-_i, n^+_i, {}_{\bar{r}_{i,N-1}}\delta_i, {}_{\bar{r}_{i,N-1}}G_i, {}_{\bar{r}_{i,N-1}}\varepsilon_i)$-expansion for the stationary probability $\pi_i(\varepsilon)$ given in Theorem 4.4 and the $(n^-_{\mathbb{U}}, n^+_{\mathbb{U}}, {}_{\bar{r}_{i,N-1}}\delta_{\mathbb{U}}, {}_{\bar{r}_{i,N-1}}G_{\mathbb{U}}, {}_{\bar{r}_{i,N-1}}\varepsilon_{\mathbb{U}})$-expansion for function $\Pi_{\mathbb{U}}(\varepsilon)$ should be used. This yields the corresponding pivotal $(n^-_{\mathbb{U},i}, n^+_{\mathbb{U},i})$-expansion for the conditional quasi-stationary stationary probability $\pi_{\mathbb{U},i}(\varepsilon)$, given in Theorem 4.5, and proves that this expansion is a $(n^-_{\mathbb{U},i}, n^+_{\mathbb{U},i}, {}_{\bar{r}_{i,N-1}}\delta_{\mathbb{U},i}, {}_{\bar{r}_{i,N-1}}G_{\mathbb{U},i}, {}_{\bar{r}_{i,N-1}}\varepsilon_{\mathbb{U},i})$-expansion, with parameters computed in the process of realization of the above algorithm. □

Remark 4.5. By proposition **(iii)** of Lemma 2.7, inequality ${}_{\bar{r}_{i,N-1}}\delta_{\mathbb{U},i} \geq \delta^*[1]$ holds, for every sequence $\bar{r}_{i,N-1}$. This makes it possible to rewrite function $\pi_{\mathbb{U},i}(\varepsilon)$ as the pivotal $(n^-_{\mathbb{U},i}, n^+_{\mathbb{U},i}, \delta^*[1], {}_{\bar{r}_{i,N-1}}G^*_{\mathbb{U},i}, {}_{\bar{r}_{i,N-1}}\varepsilon_{\mathbb{U},i})$-expansion, with parameter ${}_{\bar{r}_{i,N-1}}G^*_{\mathbb{U},i} = {}_{\bar{r}_{i,N-1}}G_{\mathbb{U},i} \cdot ({}_{\bar{r}_{i,N-1}}\varepsilon_{\mathbb{U},i})^{({}_{\bar{r}_{i,N-1}}\delta_{\mathbb{U},i} - \delta^*[1])}$.

We also would like to comment the use of the term "conditional quasi-stationary probability" for quantities defined in relations (4.8).

Let assume, for the moment, that the semi-Markov process $\eta_\varepsilon(t)$ is non-periodic (this means that the distributions of return time $\mathsf{P}_i\{\tau_i \leq t\}$ has a non-arithmetic distribution, for some, and, therefore, for any $i \in \mathbb{X}$). In this case, under conditions **A**, **B**, and $\mathbf{C}_1$, the following individual ergodic relation takes place:

$$\mathsf{P}_i\{\eta_\varepsilon(t) = j\} \to \pi_j(\varepsilon) \text{ as } t \to \infty, \text{ for } i, j \in \mathbb{X}. \tag{4.11}$$

In this case, the conditional quasi-stationary probabilities $\pi_{\mathbb{U},i}(\varepsilon), i \in \mathbb{U}$ can be defined by the relation alternative to (4.8),

$$\pi_{\mathbb{U},j}(\varepsilon) = \frac{\pi_i(\varepsilon)}{\sum_{j \in \mathbb{U}} \pi_j(\varepsilon)} = \lim_{t \to \infty} \mathsf{P}_i\{\eta_\varepsilon(t) = j / \eta_\varepsilon(t) \in \mathbb{U}\}, \text{ for } i, j \in \mathbb{U}. \tag{4.12}$$

While, the term "quasi-stationary probability (distribution)" is traditionally used for more complex limits,

$$q_j(\varepsilon) = \lim_{t \to \infty} \mathsf{P}_i\{\eta_\varepsilon(t) = j / \eta_\varepsilon(s) \in \mathbb{U}, 0 \leq s \leq t\}, \text{ for } i, j \in \mathbb{U}. \tag{4.13}$$

A detailed presentation of results concerned quasi-stationary distributions and comprehensive bibliographies of works in this area can be found in the books by Gyllenberg and Silvestrov (2008), Nåsell (2011), and Collet et al. (2013).

4.3.2 Asymptotic Expansions for Stationary Limits of Additive Functionals

Let us introduce the so-called stationary limits for additive functionals, which can be defined, for any real-valued function $f(i,\varepsilon)$, $i \in \mathbb{X}$, $\varepsilon \in (0,\varepsilon_0]$, by the following asymptotic relation, which takes place under conditions **A**, **B**, and $\mathbf{C}_1$, for every $\varepsilon \in (0,\varepsilon_0]$ (see, for example, Silvestrov (1980), Shurenkov (1989), and Limnios and Oprişan (2001)),

$$\bar{\mu}_\varepsilon(f,t) = \frac{1}{t}\int_0^t f(\eta_\varepsilon(s),\varepsilon)ds \xrightarrow{a.s} f(\varepsilon) = \sum_{j\in\mathbb{X}} f(j,\varepsilon)\pi_j(\varepsilon) \text{ as } t \to \infty. \quad (4.14)$$

Let us assume that the following perturbation condition, based on Laurent asymptotic expansions, holds:

H: $f(i,\varepsilon) = \sum_{l=k^-_{f,i}}^{k^+_{f,i}} f_i[l]\varepsilon^l + o_{f,i}(\varepsilon^{k^+_{f,i}})$, $\varepsilon \in (0,\varepsilon_0]$, for $i \in \mathbb{X}$, where **(a)** coefficients $f_i[l], k^-_{f,j} \leq l \leq k^+_{f,i}$ are real numbers, $f_i[k^-_{f,i}] > 0$ and $-\infty < k^-_{f,i} \leq k^+_{f,i} < \infty$ are integers, for $i \in \mathbb{X}$; **(b)** $o_{f,i}(\varepsilon^{k^+_{f,i}})/\varepsilon^{k^+_{f,i}} \to 0$ as $\varepsilon \to 0$, for $i \in \mathbb{X}$.

Remark 4.6. The asymptotic expansions in condition **H** are not assumed to be pivotal. In particular, it is possible that functions $f(i,\varepsilon) \equiv 0$ for some $i \in \mathbb{X}$.

The following theorem presents a Laurent asymptotic expansion with remainder given in standard form for stationary limits $f(\varepsilon)$.

Theorem 4.7. *Let conditions* **A**, **B**, $\mathbf{C}_1$, **D**, *and* $\mathbf{E}_1$ *hold for semi-Markov processes* $\eta_\varepsilon(t)$ *and condition* **H** *holds for function* $f(i,\varepsilon)$. *Then, the* (n^-_f, n^+_f)*-expansion for the stationary limit* $f(\varepsilon)$ *is given by the algorithm described below, in the proof of the theorem.*

Proof. The above algorithm includes two steps.

First, proposition **(iii)** (the multiplication rule) of Lemmas 2.2 should be applied, for every $i \in \mathbb{X}$, to asymptotic expansions for stationary probabilities $\pi_i(\varepsilon)$, given in Theorem 4.4, and function $f(i,\varepsilon)$, given in condition **H**. This yields the $(n^-_{f,i}, n^+_{f,i})$-expansion for the product $f(i,\varepsilon)\pi_i(\varepsilon)$.

Second, the proposition **(i)** (the multiple summation rule) of Lemmas 2.3 should be applied to sum $f(\varepsilon) = \sum_{i\in\mathbb{X}} f(i,\varepsilon)\pi_i(\varepsilon)$ that yields the corresponding (n^-_f, n^+_f)-expansion for function $f(\varepsilon)$. □

Let us assume that the following stronger than **H** perturbation condition holds:

H$'$: $f(i,\varepsilon) = \sum_{l=k^-_{f,i}}^{k^+_{f,i}} f_i[l]\varepsilon^l + o_{f,i}(\varepsilon^{k^+_{f,i}})$, $\varepsilon \in (0,\varepsilon_0]$, for $i \in \mathbb{X}$, where **(a)** coefficients $f_i[l], k^-_{f,j} \leq l \leq k^+_{f,i}$ are real numbers, $f_i[k^-_{f,i}] > 0$ and $-\infty < k^-_{f,i} \leq k^+_{f,i} < \infty$ are integers, for $i \in \mathbb{X}$; **(b)** $|o_{f,i}(\varepsilon^{k^+_{f,i}})| \leq G_{f,i}\varepsilon^{k^+_{f,i}+\delta_{f,i}}, \varepsilon \in (0,\varepsilon_{f,i}]$, for $i \in \mathbb{X}$, where $0 < \delta_{f,i} \leq 1, 0 \leq G_{f,i} < \infty, 0 < \varepsilon_{f,i} \leq \varepsilon_0$, for $i \in \mathbb{X}$.

The following theorem presents Laurent asymptotic expansions with explicit upper bounds for remainders for stationary limits $f(\varepsilon)$.

Theorem 4.8. *Let conditions* **A**, **B**, **C**$_1$, **D**$'$, *and* **E**$'_1$ *hold for semi-Markov processes* $\eta_\varepsilon(t)$ *and condition* **H**$'$ *holds for function* $f(i,\varepsilon)$. *Then, the* (n^-_f, n^+_f)*-expansion for the stationary limit* $f(\varepsilon)$ *given in Theorem 4.7 is a* $(n^-_f, n^+_f, {}_{\bar{r}_{i,N-1}}\delta_f, {}_{\bar{r}_{i,N-1}}G_f, {}_{\bar{r}_{i,N-1}}\varepsilon_f)$*-expansion. Parameters* ${}_{\bar{r}_{i,N-1}}\delta_f$, ${}_{\bar{r}_{i,N-1}}G_f$, *and* ${}_{\bar{r}_{i,N-1}}\varepsilon_f$ *can be computed using the algorithm described below, in the proof of the theorem.*

Proof. The above algorithm includes two steps.

First, proposition **(iii)** (the multiplication rule) of Lemmas 2.2 and 2.6 should be applied, for every $i \in \mathbb{X}$, to asymptotic expansion the product $f(i,\varepsilon)\pi_j(\varepsilon)$ given in proof of Theorem 4.7. This yields the $(n^-_{f,i}, n^+_{f,i}, {}_{\bar{r}_{i,N-1}}\delta_{f,i}, {}_{\bar{r}_{i,N-1}}G_{f,i}, {}_{\bar{r}_{i,N-1}}\varepsilon_{f,i})$-expansion for the product $f(i,\varepsilon)\pi_i(\varepsilon)$.

Second, the proposition **(i)** (the multiple summation rule) of Lemmas 2.3 and 2.7 should be applied to asymptotic expansion for sum $f(\varepsilon) = \sum_{i\in\mathbb{X}} f(i,\varepsilon)\pi_i(\varepsilon)$ given in Theorem 4.7 that yields the corresponding $(n^-_f, n^+_f, {}_{\bar{r}_{i,N-1}}\delta_f, {}_{\bar{r}_{i,N-1}}G_f, {}_{\bar{r}_{i,N-1}}\varepsilon_f)$-expansion for function $f(\varepsilon)$. □

Chapter 5
Nonlinearly Perturbed Birth-Death-Type Semi-Markov Processes

In Chapter 5, we present asymptotic expansions for stationary and conditional quasi-stationary distributions of nonlinearly perturbed birth-death-type semi-Markov processes. In this case, the corresponding expansions can be given in a more explicit form.

We refer to books by Karlin and Taylor (1975), Van Doorn (1981), Kovalenko et al. (1996), Pinsky and Karlin (2011), where one can found the basic facts about birth-death-type stochastic processes.

It is well known from the literature that explicit formulas for birth-death models are bulky. This is well seen from the explicit formulas for stationary distributions of birth-death-type semi-Markov processes given in Lemma 5.1. These formulas are, indeed, complex rational functions of transition probabilities for the corresponding embedded Markov chain and expectations of sojourn times.

The algorithm for construction of asymptotic expansions, with remainders given in the standard form of $o(\cdot)$, for stationary probabilities of nonlinearly perturbed birth-death-type semi-Markov processes is described in Lemmas 5.2–5.6 and Theorem 5.1. It is based on sequential application of the corresponding operational rules for Laurent asymptotic expansions given in Section 2.1 to functions appearing in the above lemmas and theorem.

Analogously, the algorithm for construction of asymptotic expansions, with explicit upper bounds for remainders, for stationary probabilities of nonlinearly perturbed birth-death-type semi-Markov processes is described in Lemmas 5.8–5.12 and Theorem 5.3. It is based on sequential application of the corresponding operational rules for Laurent asymptotic expansions with explicit upper bounds for remainders given in Section 2.2 to functions appearing in the above lemmas and theorem.

Also, additional algorithms required for construction of asymptotic expansions, with and without explicit upper bounds for remainders, for conditional quasi-stationary distributions of nonlinearly perturbed birth-death-type semi-Markov processes are described, respectively, in Lemma 5.5 and Theorem 5.2, and in Lemma 5.13 and Theorem 5.4.

D. Silvestrov, S. Silvestrov, *Nonlinearly Perturbed Semi-Markov Processes*,
SpringerBriefs in Probability and Mathematical Statistics,
DOI 10.1007/978-3-319-60988-1_5

The visual complexity of the corresponding recurrent formulas is not an obstacle for effective computations. Moreover, the corresponding recurrent formulas are well prepared for their programming.

5.1 Stationary and Quasi-Stationary Distributions for Perturbed Birth-Death-Type Semi-Markov Processes

In this section, we introduce a model of perturbed birth-death-type semi-Markov processes, define stationary and conditional quasi-stationary distributions for such processes, and formulate basic perturbation conditions.

5.1.1 Perturbed Birth-Death-Type Semi-Markov Processes

Let us $\eta_\varepsilon(t), t \geq 0$ be, for every $\varepsilon \in (0, \varepsilon_0]$, a semi-Markov process with a phase space $\mathbb{X} = \{0, \ldots, N\}$ and transition probabilities $Q_{\varepsilon,ij}(t), t \geq 0, i, j \in \mathbb{X}$. We assume that $N \geq 1$.

This process has the birth-death type if the following relation holds, for its transition probabilities, for $t \geq 0$,

$$Q_{\varepsilon,ij}(t) = \begin{cases} F_{\varepsilon,0,\pm}(t)p_{0,\pm}(\varepsilon) & \text{if } j = 0 + \frac{1\pm1}{2}, \text{ for } i = 0, \\ F_{\varepsilon,i,\pm}(t)p_{i,\pm}(\varepsilon) & \text{if } j = i \pm 1, \quad \text{for } 0 < i < N, \\ F_{\varepsilon,N,\pm}(t)p_{N,\pm}(\varepsilon) & \text{if } j = N - \frac{1\mp1}{2}, \text{ for } i = N, \\ 0 & \text{otherwise,} \end{cases} \tag{5.1}$$

where (a) $F_{\varepsilon,i,\pm}(t), t \geq 0$ is for every $i \in \mathbb{X}$ and $\varepsilon \in (0, \varepsilon_0]$, a distribution function concentrated on $[0, \infty)$; (b) $p_{i,\pm}(\varepsilon) \geq 0, p_{i,-}(\varepsilon) + p_{i,+}(\varepsilon) = 1$, for every $i \in \mathbb{X}$ and $\varepsilon \in (0, \varepsilon_0]$.

Condition **A** takes in this case the following form:

J: $p_{i,\pm}(\varepsilon) > 0$, for $0 \leq i \leq N$, for $\varepsilon \in (0, \varepsilon_0]$.

Condition **B** takes, in this case, the following form:

K: $F^{(\varepsilon)}_{i,\pm}(0) = 0, i \in \mathbb{X}$, for $\varepsilon \in (0, \varepsilon_0]$.

Let us also denote, for $i, j \in \mathbb{X}$,

$$f_{i,\pm}(\varepsilon) = \int_0^\infty t F^{(\varepsilon)}_{i,\pm}(dt), \; e_{i,\pm}(\varepsilon) = f_{i,\pm}(\varepsilon) p_{i,\pm}(\varepsilon). \tag{5.2}$$

We also assume that condition $\mathbf{C}_1$ holds. It takes, in this case, the following form:

L: $f_{i,\pm}(\varepsilon) < \infty, i \in \mathbb{X}$, for $\varepsilon \in (0, \varepsilon_0]$.

The following relations take place:

$$p_{ij}(\varepsilon) = \begin{cases} p_{0,\pm}(\varepsilon) & \text{if } j = 0 + \frac{1\pm 1}{2}, & \text{for } i = 0, \\ p_{i,\pm}(\varepsilon) & \text{if } j = i \pm 1, & \text{for } 0 < i < N, \\ p_{N,\pm}(\varepsilon) & \text{if } j = N - \frac{1\mp 1}{2}, & \text{for } i = N, \\ 0 & \text{otherwise}, \end{cases} \tag{5.3}$$

and

$$e_{ij}(\varepsilon) = \begin{cases} e_{0,\pm}(\varepsilon) & \text{if } j = 0 + \frac{1\pm 1}{2}, & \text{for } i = 0, \\ e_{i,\pm}(\varepsilon) & \text{if } j = i \pm 1, & \text{for } 0 < i < N, \\ e_{N,\pm}(\varepsilon) & \text{if } j = N - \frac{1\mp 1}{2}, & \text{for } i = N, \\ 0 & \text{otherwise}. \end{cases} \tag{5.4}$$

Let us assume that there exists some integer $0 \leq L < \infty$ such that the following perturbation conditions hold:

$\mathbf{M}_L$: $p_{i,\pm}(\varepsilon) = \sum_{l=l_{i,\pm}}^{L+l_{i,\pm}} a_{i,\pm}[l]\varepsilon^l + o_{i,\pm}(\varepsilon^{L+l_{i,\pm}})$, $\varepsilon \in (0, \varepsilon_0]$, for $i \in \mathbb{X}$, where: **(a)** coefficients $a_{i,\pm}[l], l_{i,\pm} \leq l \leq L + l_{i,\pm}$ are real numbers, $a_{i,\pm}[l_{i,\pm}] > 0$, for $i \in \mathbb{X}$; **(b)** $l_{i,\pm} = 0$, for $0 < i < N$; $l_{0,-} = 0$, $l_{0,+} = 0$ or $l_{0,+} = 1$; and $l_{N,+} = 0$, $l_{N,-} = 0$ or $l_{N,-} = 1$; **(c)** $o_{i,\pm}(\varepsilon^{L+l_{i,\pm}})/\varepsilon^{L+l_{i,\pm}} \to 0$ as $\varepsilon \to 0$, for $i \in \mathbb{X}$.

and

$\mathbf{N}_L$: $e_{i,\pm}(\varepsilon) = \sum_{l=l_{i,\pm}}^{L+l_{i,\pm}} b_{i,\pm}[l]\varepsilon^l + \dot{o}_{i,\pm}(\varepsilon^{L+l_{i,\pm}})$, $\varepsilon \in (0, \varepsilon_0]$, for $i \in \mathbb{X}$, where: **(a)** coefficients $b_{i,\pm}[l], l_{i,\pm} \leq l \leq L + l_{i,\pm}$ are real numbers, $b_{i,\pm}[l_{i,\pm}] > 0$, for $i \in \mathbb{X}$; **(b)** $\dot{o}_{i,\pm}(\varepsilon^{L+l_{i,\pm}})/\varepsilon^{L+l_{i,\pm}} \to 0$ as $\varepsilon \to 0$, for $i \in \mathbb{X}$.

It is worth noting that the same parameters L and $l_{i,\pm}, i \in \mathbb{X}$ appear in conditions $\mathbf{M}_L$ and $\mathbf{N}_L$ that makes consistent the asymptotic expansions in these conditions.

In the simplest case, $L = 0$, the above conditions take the following forms:

$\mathbf{M}_0$: $p_{i,\pm}(\varepsilon) = a_{i,\pm}[l_{i,\pm}]\varepsilon^{l_{i,\pm}} + o_{i,\pm}(\varepsilon^{l_{i,\pm}})$, $\varepsilon \in (0, \varepsilon_0]$, for $i \in \mathbb{X}$, where: **(a)** $a_{i,\pm}[l_{i,\pm}] > 0$, for $i \in \mathbb{X}$; **(b)** $l_{i,\pm} = 0$, for $0 < i < N$; $l_{0,-} = 0$, $l_{0,+} = 0$ or $l_{0,+} = 1$; and $l_{N,+} = 0$, $l_{N,-} = 0$ or $l_{N,-} = 1$; **(c)** $o_{i,\pm}(\varepsilon^{l_{i,\pm}})/\varepsilon^{l_{i,\pm}} \to 0$ as $\varepsilon \to 0$, for $i \in \mathbb{X}$.

and

$\mathbf{N}_0$: $e_{i,\pm}(\varepsilon) = b_{i,\pm}[l_{i,\pm}]\varepsilon^{l_{i,\pm}} + \dot{o}_{i,\pm}(\varepsilon^{l_{i,\pm}})$, $\varepsilon \in (0, \varepsilon_0]$, for $i \in \mathbb{X}$, where: **(a)** $b_{i,\pm}[l_{i,\pm}] > 0$, for $i \in \mathbb{X}$; **(b)** $\dot{o}_{i,\pm}(\varepsilon^{l_{i,\pm}})/\varepsilon^{l_{i,\pm}} \to 0$ as $\varepsilon \to 0$, for $i \in \mathbb{X}$.

It is useful to explain what role is played by parameter l_i in conditions $\mathbf{M}_L$ and $\mathbf{N}_L$. This parameter equalizes the length of asymptotic expansions appearing in these conditions. All expansions appearing in conditions $\mathbf{M}_L$ and $\mathbf{N}_L$ have the length L.

Note that conditions $\mathbf{M}_L$ and $\mathbf{N}_L$ imply that there exist $\varepsilon'_0 \in (0, \varepsilon_0]$ such that probabilities $p_{i,\pm}(\varepsilon) > 0, i \in \mathbb{X}$ and expectations $e_{i,\pm}(\varepsilon) > 0, i \in \mathbb{X}$ for $\varepsilon \in (0, \varepsilon'_0]$. This let us, just, assume that $\varepsilon'_0 = \varepsilon_0$.

Condition $\mathbf{M}_L$ implies that there exist limits, $\lim_{\varepsilon \to 0} p_{i,\pm}(\varepsilon) = p_{i,\pm}(0)$, for $i \in \mathbb{X}$. These limits are: (a) $p_{i,\pm}(0) = a_{i,\pm}[0] > 0$, for $0 < i < N$; (b) $p_{0,-}(0) = a_{0,-}[0] > 0$; (c) $p_{0,+}(0) = a_{0,+}[0] > 0$ if $l_{0,+} = 0$ or 0 if $l_{0,+} = 1$; (d) $p_{N,+}(0) = a_{N,+}[0] > 0$; (e) $p_{N,-}(0) = a_{N,-}[0] > 0$ if $l_{N,-} = 0$ or 0 if $l_{N,-} = 1$.

Process $\eta_\varepsilon(t)$ has no asymptotically absorbing states, if $p_{0,+}(0), p_{N,-}(0) > 0$, i.e., $l_{0,+}, l_{N,-} = 1$. The above relations also imply that state 0 is asymptotically absorbing state for process $\eta_\varepsilon(t)$, if $p_{0,+}(0) = 0$, i.e., $l_{0,+} = 0$. Analogously, state N is asymptotically absorbing state for process $\eta_\varepsilon(t)$, if $p_{N,-}(0) = 0$, i.e., $l_{N,-} = 0$.

The model assumption, $p_{i,-}(\varepsilon) + p_{i,+}(\varepsilon) = 1, \varepsilon \in (0, \varepsilon_0]$, also implies that the following condition should automatically hold:

$\mathbf{O}_L$: **(a)** $a_{i,-}[0] + a_{i,+}[0] = 1, a_{i,-}[l] + a_{i,+}[l] = 0, 1 \leq l \leq L$ and $o_{i,-}(\varepsilon^L) + o_{i,+}(\varepsilon^L) = 0, \varepsilon \in (0, \varepsilon_0]$, for $0 < i < N$; **(b)** $a_{0,-}[0] + a_{0,+}[0] = 1, a_{0,-}[l] + a_{0,+}[l] = 0, 1 \leq l \leq L$ and $o_{0,-}(\varepsilon^L) + o_{0,+}(\varepsilon^L) = 0, \varepsilon \in (0, \varepsilon_0]$, if $l_{0,+} = 0$; or $a_{0,-}[0] = 1, a_{0,-}[l] + a_{0,+}[l] = 0, 1 \leq l \leq L$ and $o_{0,-}(\varepsilon^L) + a_{0,+}[L+1]\varepsilon^{L+1} + o_{0,+}(\varepsilon^{L+1}) = 0, \varepsilon \in (0, \varepsilon_0]$, if $l_{0,+} = 1$; **(c)** $a_{N,-}[0] + a_{N,+}[0] = 1, a_{N,-}[l] + a_{N,+}[l] = 0, 1 \leq l \leq L$ and $o_{N,-}(\varepsilon^L) + o_{N,+}(\varepsilon^L) = 0, \varepsilon \in (0, \varepsilon_0]$, if $l_{N,-} = 0$; or $a_{N,+}[0] = 1, a_{N,-}[l] + a_{N,+}[l] = 0, 1 \leq l \leq L$ and $a_{N,-}[L+1]\varepsilon^{L+1} + o_{N,-}(\varepsilon^{L+1}) + o_{N,+}(\varepsilon^L) = 0, \varepsilon \in (0, \varepsilon_0]$, if $l_{N,-} = 1$.

Condition $\mathbf{N}_L$ implies that there exist limits, $\lim_{\varepsilon \to 0} e_{i,\pm}(\varepsilon) = e_{i,\pm}(0)$, for $i \in \mathbb{X}$. These limits are: (a) $e_{i,\pm}(0) = b_{i,\pm}[0] > 0$, for $0 < i < N$; (b) $e_{0,-}(0) = b_{0,-}[0] > 0$; (c) $e_{0,+}(0) = b_{0,+}[0] > 0$ if $l_{0,+} = 0$ or 0 if $l_{0,+} = 1$; (d) $e_{N,+}(0) = b_{N,+}[0] > 0$; (e) $e_{N,-}(0) = b_{N,-}[0] > 0$ if $l_{N,-} = 0$ or 0 if $l_{N,-} = 1$.

Thus, conditions $\mathbf{M}_L$ and $\mathbf{N}_L$ also imply that the following consistency condition holds:

P: $e_{i,\pm}(0) > 0$ if and only if $p_{i,\pm}(0) > 0$, for $i = 0, N$.

In the case, where asymptotic expansions with explicit upper bounds for remainders are objects of interest, perturbation conditions $\mathbf{M}_L$ and $\mathbf{N}_L$ should be replaced by stronger conditions, respectively;

$\mathbf{M}'_L$: $p_{i,\pm}(\varepsilon) = \sum_{l=l_{i,\pm}}^{L+l_{i,\pm}} a_{i,\pm}[l]\varepsilon^l + o_{i,\pm}(\varepsilon^{L+l_{i,\pm}})$, $\varepsilon \in (0, \varepsilon_0]$, for $i \in \mathbb{X}$, where: **(a)** coefficients $a_{i,\pm}[l], l_{i,\pm} \leq l \leq L + l_{i,\pm}$ are real numbers, $a_{i,\pm}[l_{i,\pm}] > 0$, for $i \in \mathbb{X}$; **(b)** $l_{i,\pm} = 0$, for $0 < i < N$; $l_{0,-} = 0$, $l_{0,+} = 0$ or $l_{0,+} = 1$; and $l_{N,+} = 0$, $l_{N,-} = 0$ or $l_{N,-} = 1$; **(c)** $|o_{i,\pm}(\varepsilon^{L+l_{i,\pm}})| \leq G_{i,\pm}\varepsilon^{L+l_{i,\pm}+\delta_{i,\pm}}$, $\varepsilon \in (0, \varepsilon_{i,\pm}]$, for $i \in \mathbb{X}$, where: $0 < \delta_{i,\pm} \leq 1, 0 \leq G_{i,\pm} < \infty, 0 < \varepsilon_{i,\pm} \leq \varepsilon_0$, for $i \in \mathbb{X}$.

and

$\mathbf{N}'_L$: $e_{i,\pm}(\varepsilon) = \sum_{l=l_{i,\pm}}^{L+l_{i,\pm}} b_{i,\pm}[l]\varepsilon^l + \dot{o}_{i,\pm}(\varepsilon^{L+l_{i,\pm}})$, $\varepsilon \in (0, \varepsilon_0]$, for $i \in \mathbb{X}$, where: **(a)** coefficients $b_{i,\pm}[l], l_{i,\pm} \leq l \leq L + l_{i,\pm}$ are real numbers, $b_{i,\pm}[l_{i,\pm}] > 0$, for $i \in \mathbb{X}$; **(c)** $|\dot{o}_{i,\pm}(\varepsilon^{L+l_{i,\pm}})| \leq \dot{G}_{i,\pm}\varepsilon^{L+l_{i,\pm}+\dot{\delta}_{i,\pm}}$, $\varepsilon \in (0, \dot{\varepsilon}_{i,\pm}]$, for $i \in \mathbb{X}$, where: $0 < \dot{\delta}_{i,\pm} \leq 1, 0 \leq \dot{G}_{i,\pm} < \infty, 0 < \dot{\varepsilon}_{i,\pm} \leq \varepsilon_0$, for $i \in \mathbb{X}$.

Conditions $\mathbf{M}'_L$ and $\mathbf{N}'_L$ differ from conditions $\mathbf{M}_L$ and $\mathbf{N}_L$ by the assumptions imposed on the remainders of asymptotic expansions appearing in these conditions. In conditions $\mathbf{M}_L$ and $\mathbf{N}_L$, the remainders are given in the standard form of $o(\cdot)$. In conditions $\mathbf{M}'_L$ and $\mathbf{N}'_L$ the remainders are given in the form with explicit upper bounds.

We use periods above the letters denoting parameters and remainders of asymptotic expansions, which appear in conditions $\mathbf{N}_d$ and $\mathbf{N}'_d$, in order to distingue them of parameters and remainders of asymptotic expansions, which appear in conditions $\mathbf{M}_L$ and $\mathbf{M}'_L$.

Since condition $\mathbf{M}'_L$ is stronger than condition $\mathbf{M}_L$, condition $\mathbf{O}_L$ also holds.

5.1.2 Stationary and Quasi-Stationary Distributions for Perturbed Birth-Death-Type Semi-Markov Processes

There are three basic variants of the model, where one of the following conditions hold:

$\mathbf{Q}_1$: $p_{0,+}(0), p_{N,-}(0) > 0$, i.e., $l_{0,+}, l_{N,-} = 0$.

$\mathbf{Q}_2$: $p_{0,+}(0) = 0, p_{N,-}(0) > 0$, i.e., $l_{0,+} = 1, l_{N,-} = 0$.

$\mathbf{Q}_3$: $p_{0,+}(0), p_{N,-}(0) = 0$, i.e., $l_{0,+}, l_{N,-} = 1$.

The case, $p_{0,+}(0) > 0, p_{N,-}(0) = 0$, is analogous to the case, where condition $\mathbf{Q}_2$ holds, and we omit its consideration.

The limiting birth-death-type Markov chain $\eta_{0,n}$ with the matrix of transition probabilities $\|p_{ij}(0)\|$ has: **(a)** one class of communicative states $\mathbb{U}_1 = \mathbb{X}$, if condition $\mathbf{Q}_1$ holds, **(b)** one communicative class of transient states $\mathbb{U}_2 = {}_0\mathbb{X} = \mathbb{X} \setminus \{0\}$ and the absorbing state 0, if condition $\mathbf{Q}_2$ holds, and **(c)** one communicative class of transient states $\mathbb{U}_3 = {}_{0,N}\mathbb{X} = \mathbb{X} \setminus \{0, N\}$ and two absorbing states 0 and N, if condition $\mathbf{Q}_3$ holds.

Conditions **J–L** imply that the birth-death-type semi-Markov process $\eta_\varepsilon(t)$ is, for every $\varepsilon \in (0, \varepsilon_0]$, ergodic. Let $\pi_i(\varepsilon), i \in \mathbb{X}$ be its stationary distribution. Conditions **J–L** and $\mathbf{M}_L$, $\mathbf{N}_L$ imply that stationary probabilities $\pi_i(\varepsilon)$ converge as $\varepsilon \to 0$, i.e., there exist limits $\lim_{\varepsilon \to 0} \pi_i(\varepsilon) = \pi_i(0)$ as $\varepsilon \to 0$, for $i \in \mathbb{X}$.

Moreover, we shall show that the limiting stationary probabilities $\pi_i(0) > 0, i \in \mathbb{X}$, if condition $\mathbf{Q}_1$ holds; $\pi_0(0) = 1, \pi_i(0) = 0, i \in {}_0\mathbb{X}$, if condition $\mathbf{Q}_2$ holds; and $\pi_0(0), \pi_N(0) > 0, \pi_0(0) + \pi_N(0) = 1, \pi_i(0) = 0, i \in {}_{0,N}\mathbb{X}$, if condition $\mathbf{Q}_3$ holds.

We are also interested in conditional quasi-stationary probabilities, for $r = 1, 2, 3$,

$$\pi_{\mathbb{U}_r,i}(\varepsilon) = \frac{\pi_i(\varepsilon)}{\sum_{j \in \mathbb{U}_r} \pi_j(\varepsilon)}, \ i \in \mathbb{U}_r. \tag{5.5}$$

Our goal is to get, under conditions **J–L**, $\mathbf{M}_L$–$\mathbf{O}_L$, **P**, and $\mathbf{Q}_r$ (for $r = 1, 2, 3$), asymptotic expansions for the stationary probabilities $\pi_i(\varepsilon), i \in \mathbb{X}$ and for the conditional quasi-stationary probabilities $\pi_{\mathbb{U}_r,i}(\varepsilon), i \in \mathbb{U}_r$.

In the case $r = 1$, the conditional quasi-stationary probabilities $\pi_{\mathbb{U}_r,i}(\varepsilon) = \pi_i(\varepsilon)$, $i \in \mathbb{U}_1 = \mathbb{X}$. Thus, the only cases $r = 2$ and $r = 3$ do require some special asymptotic analysis for the conditional quasi-stationary probabilities. Moreover, we assume that $N \geq 2$ if $r = 3$, in order to escape the case of empty set $\mathbb{U}_3$.

It is worth to note that the transition probabilities $p_{0,-}(\varepsilon)$ and $p_{N,+}(\varepsilon)$ are, actually, involved neither in the explicit formulas for stationary probabilities $\pi_i(\varepsilon), 0 \leq i \leq N$ nor in the algorithms for construction of the corresponding asymptotic expansions for these probabilities described below. This let one consider alternatives to the assumptions imposed on probabilities $p_{0,-}(\varepsilon)$ and $p_{N,+}(\varepsilon)$ in conditions $\mathbf{M}_L$ and $\mathbf{M}'_L$, where it is assumed that these probabilities can be expanded in pivotal $(0,L)$-expansions. Alternatively, it can be, just, assumed that one or both probabilities $p_{0,-}(\varepsilon)$ and $p_{N,+}(\varepsilon)$ can be expanded, respectively, in pivotal (l_0,l_0)- and (l_N,l_N) expansions, for some integer parameter $l_0, l_N \geq 1$. Moreover, an alternative variant of condition $\mathbf{J}$ can be considered, where one or both probabilities $p_{0,-}(\varepsilon)$ and $p_{N,+}(\varepsilon)$ identically equal to 0.

The above alternatives can be actual, for probability $p_{0,-}(\varepsilon)$, if condition $\mathbf{Q}_1$ holds, and, for probability $p_{N,+}(\varepsilon)$, if condition $\mathbf{Q}_1$ or $\mathbf{Q}_2$ holds.

Neither formulas for stationary probabilities $\pi_i(\varepsilon), 0 \leq i \leq N$ nor the algorithms for construction of the corresponding asymptotic expansions for these probabilities change.

One of the well-known interpretations for the birth-death-type semi-Markov process $\eta_\varepsilon(t)$ (at least for two particular cases, where this process is a discrete or continuous time Markov chain) is to consider it as a process which describes stochastic dynamics of a finite population and represents its size at instant t. In this case, the transition probabilities $p_{i,+}(\varepsilon)$ and $p_{i,-}(\varepsilon)$ are, respectively, birth and death probabilities (during one transition period) for the case, where the size of population is $0 < i < N$. The size of population cannot exceed N. If the birth of a new individual occurs, when the population has the maximal size N, one individual leaves the population. Probability $p_{N,+}(\varepsilon)$ can be interpreted as a probability of emigration (during one transition period). After an extinction of population, when its size becomes equal 0, the population can recover due to immigration of some individual. The probability $p_{0,+}(\varepsilon)$ can be interpreted as a probability of immigration (during one transition period). Both variants of the model, where condition $\mathbf{Q}_1$ or $\mathbf{Q}_2$ holds, are actual.

Another well-known interpretation of process $\eta_\varepsilon(t)$ is to consider it as a process which describes stochastic dynamics of an epidemic model and represents, for example, the number of contaminated individuals at instant t in a population of size N. In this case, the transition probabilities $p_{i,+}(\varepsilon)$ and $p_{i,-}(\varepsilon)$ are, respectively, contamination and recovering probabilities (during one transition period) for the case, where the number of contaminated individuals is $0 < i < N$. Probability $p_{0,+}(\varepsilon)$ can be interpreted as a probability of contamination via external contacts of individuals with infected ones outside of the population. Probability $p_{N,+}(\varepsilon)$ can be interpreted as a recovering probability (during one transition period) for the case where all individuals in the population are contaminated. Again, both variants of the model, where condition $\mathbf{Q}_1$ or $\mathbf{Q}_2$ holds, are actual.

One of the models of population genetics relates to one-sex finite population of a size N. Individuals from this population carry two copies of a certain gene, which can exist in two variants (alleles), say, of type A and B. The birth-death-type process $\eta_\varepsilon(t)$ describes stochastic dynamics of the number of individuals in the population which carries gene of type A at instant t. In this case, transition probabilities $p_{i,+}(\varepsilon)$ and $p_{i,-}(\varepsilon)$ are, respectively, probabilities of increase or decrease by 1 the number of individuals carrying gene of type A (during one transition period), for the case, where the number of such individual is $0 < i < N$. Probabilities $p_{0,+}(\varepsilon)$ and $p_{N,-}(\varepsilon)$ can be interpreted as probabilities of increase by 1 and decrease by 1 the number of individuals carrying gene of type A (during one transition period), for the case when, the number of such individuals in the population is, respectively, 0 or N. This can happen due to action of some mutation factor and such probabilities are usually small. In this case, the variant of model, where condition $\mathbf{Q}_3$ holds, is actual. Some additional details concerned this model can be found in Subsection 6.3.3.

The detailed analysis of above perturbed stochastic models of population dynamics, epidemic models, and models of mathematical genetics is presented in the recent paper by Silvestrov et al. (2016).

5.1.3 Reduced Birth-Death-Type Semi-Markov Processes

Let us, first, consider the case, where the state 0 is excluded from the phase space $\mathbb{X}$. In this case, the reduced phase space ${}_0\mathbb{X} = \{1,\dots,N\}$.

We assume that the initial distribution of the semi-Markov process $\eta_\varepsilon(t)$ is concentrated on the reduced phase space ${}_0\mathbb{X}$.

The transition probabilities of the reduced birth-death-type semi-Markov process ${}_0\eta_\varepsilon(t)$ has, for every $\varepsilon \in (0,\varepsilon_0]$, the following form, for $t \geq 0$:

$$
{}_0Q_{\varepsilon,ij}(t) = \begin{cases} F_{\varepsilon,1,+}(t)p_{1,+}(\varepsilon) & \text{if } j = 2, i = 1, \\ {}_0F_{\varepsilon,1,-}(t)p_{1,-}(\varepsilon) & \text{if } j = 1, i = 1, \\ F_{\varepsilon,i,\pm}(t)p_{i,\pm}(\varepsilon) & \text{if } j = i \pm 1, 1 < i < N, \\ F_{\varepsilon,N,\pm}(t)p_{N,\pm}(\varepsilon) & \text{if } j = N - \frac{1\mp 1}{2}, i = N, \\ 0 & \text{otherwise}, \end{cases} \tag{5.6}
$$

where

$$
{}_0F_{\varepsilon,1,-}(t) = \sum_{n=0}^{\infty} F_{\varepsilon,1,-}(t) * F^{*n}_{\varepsilon,0,-}(t) * F_{\varepsilon,0,+}(t) \cdot p_{0,-}(\varepsilon)^n p_{0,+}(\varepsilon). \tag{5.7}
$$

This relation implies, for every $\varepsilon \in (0,\varepsilon_0]$, the following relation for transition probabilities of the reduced embedded Markov chain ${}_0\eta_{\varepsilon,n}$,

$$
{}_0p_{ij}(\varepsilon) = \begin{cases} {}_0p_{1,\pm}(\varepsilon) = p_{1,\pm}(\varepsilon) & \text{if } j = 1 + \frac{1\pm 1}{2}, i = 1, \\ {}_0p_{i,+}(\varepsilon) = p_{i,\pm}(\varepsilon) & \text{if } j = i \pm 1, 1 < i < N, \\ {}_0p_{N,+}(\varepsilon) = p_{N,\pm}(\varepsilon) & \text{if } j = N - \frac{1\mp 1}{2}, i = N, \\ 0 & \text{otherwise}, \end{cases} \tag{5.8}
$$

and the following relation for expectations of transition times for the reduced semi-Markov process ${}_0\eta_\varepsilon(t)$,

$$
{}_0e_{ij}(\varepsilon) = \begin{cases} {}_0e_{1,+}(\varepsilon) = e_{1,+}(\varepsilon) & \text{if } j = 2, i = 1, \\ {}_0e_{1,-}(\varepsilon) = e_{1,-}(\varepsilon) & \\ \quad + e_0(\varepsilon) \cdot \frac{p_{1,-}(\varepsilon)}{p_{0,+}(\varepsilon)} & \text{if } j = 1, i = 1, \\ {}_0e_{i,\pm}(\varepsilon) = e_{i,\pm}(\varepsilon) & \text{if } j = i \pm 1, 1 < i < N, \\ {}_0e_{N,\pm}(\varepsilon) = e_{N,\pm}(\varepsilon) & \text{if } j = N - \frac{1 \mp 1}{2}, i = N, \\ 0 & \text{otherwise}, \end{cases} \tag{5.9}
$$

where

$$
e_i(\varepsilon) = e_{i,-}(\varepsilon) + e_{i,+}(\varepsilon), \ i \in \mathbb{X}. \tag{5.10}
$$

Note that, by Theorem 3.1, the following relation takes place, for every $\varepsilon \in (0, \varepsilon_0]$ and $i, j \in {}_0\mathbb{X}$,

$$
\mathsf{E}_i \tau_{\varepsilon,j} = \mathsf{E}_{i\,0}\tau_{\varepsilon,j}. \tag{5.11}
$$

In analogous way, state N is excluded from the phase space $\mathbb{X}$. In this case, the reduced phase space ${}_N\mathbb{X} = \{0, \ldots, N-1\}$.

It is readily seen that in both cases, where $r = 0$ or $r = N$, the reduced semi-Markov process ${}_r\eta_\varepsilon(t)$ also has a birth-death type, with the phase space, respectively, ${}_0\mathbb{X} = \{1, \ldots, N\}$ or ${}_N\mathbb{X} = \{0, \ldots, N-1\}$ and transition characteristics given, respectively, by relations (5.6)–(5.9), if $r = 0$, or by their "right-hand" analogs, if $r = N$.

Let $0 \le k \le i \le r \le N$. The states $0, \ldots, k-1$ and $N, \ldots, r+1$ can be sequentially excluded from the phase space $\mathbb{X}$ of the semi-Markov process $\eta_\varepsilon(t)$.

In order to describe this recurrent procedure, let us denote the resulted reduced birth-death-type semi-Markov process as ${}_{\langle k,r \rangle}\eta_\varepsilon(t)$. This process has the reduced phase space ${}_{\langle k,r \rangle}\mathbb{X} = \{k, \ldots, r\}$.

In particular, the initial semi-Markov process $\eta_\varepsilon(t) = {}_{\langle 0,N \rangle}\eta_\varepsilon(t)$.

The reduced semi-Markov process ${}_{\langle k,r \rangle}\eta_\varepsilon(t)$ can be obtained by excluding of the state $k-1$ from the phase space ${}_{\langle k-1,j \rangle}\mathbb{X}$ of the reduced semi-Markov process ${}_{\langle k-1,r \rangle}\eta_\varepsilon(t)$ or by excluding state $r+1$ from the phase space ${}_{\langle k,r+1 \rangle}\mathbb{X}$ of the reduced semi-Markov process ${}_{\langle k,r+1 \rangle}\eta_\varepsilon(t)$.

The sequential excluding of the states $0, \ldots, k-1$, and $N, \ldots, r+1$ can be realized recurrently, by excluding the corresponding next state from arbitrary chosen one of the above two sequences.

The simplest variants for the sequences of excluded states are $0, \ldots, k-1, N, \ldots, r+1$, and $N, \ldots, r+1, 0, \ldots, k-1$.

The resulting reduced semi-Markov process ${}_{\langle k,r \rangle}\eta_\varepsilon(t)$ will be the same and it will have a birth-death type.

Here, we also accept the reduced semi-Markov process ${}_{\langle i,i \rangle}\eta_\varepsilon(t)$ with one-state phase space ${}_{\langle i,i \rangle}\mathbb{X} = \{i\}$ as a birth-death semi-Markov process.

This process has transition probability for the embedded Markov chain,

$$
{}_{\langle i,i \rangle}p_{ii}(\varepsilon) = {}_{\langle i,i \rangle}p_{i,+}(\varepsilon) + {}_{\langle i,i \rangle}p_{i,-}(\varepsilon) = 1, \tag{5.12}
$$

and the semi-Markov transition probabilities,

$$
\begin{aligned}
{}_{\langle i,i\rangle}Q_{\varepsilon,ii}(t) &= {}_{\langle i,i\rangle}F_{\varepsilon,i,+}(t)\,{}_{\langle i,i\rangle}p_{i,+}(\varepsilon) + {}_{\langle i,i\rangle}F_{\varepsilon,i,-}(t)\,{}_{\langle i,i\rangle}p_{i,-}(\varepsilon) \\
&= \mathsf{P}_i\{\tau_{\varepsilon,i} \le t\},\ t \ge 0,\ i \in \mathbb{X}.
\end{aligned} \tag{5.13}
$$

The following relations, which are, in fact, variants of relations (5.8) and (5.9) express transition probabilities ${}_{\langle k,r\rangle}p_{ij}(\varepsilon)$ and expectations of transition times ${}_{\langle k,r\rangle}e_{ij}(\varepsilon)$ for the reduced embedded semi-Markov process ${}_{\langle k,r\rangle}\eta_\varepsilon(t)$, via the transition probabilities ${}_{\langle k-1,r\rangle}p_{ij}(\varepsilon)$ and the expectations of transition times ${}_{\langle k-1,r\rangle}e_{ij}(\varepsilon)$ for the reduced embedded semi-Markov process ${}_{\langle k-1,r\rangle}\eta_\varepsilon(t)$, for $1 \le k \le r \le N$ and, for every $\varepsilon \in (0, \varepsilon_0]$,

$$
{}_{\langle k,r\rangle}p_{ij}(\varepsilon) = \begin{cases}
{}_{\langle k,r\rangle}p_{k,\pm}(\varepsilon) = {}_{\langle k-1,r\rangle}p_{k,\pm}(\varepsilon) \text{ if } j = k + \frac{1\pm1}{2}, i = k, \\
{}_{\langle k,r\rangle}p_{i,\pm}(\varepsilon) = {}_{\langle k-1,r\rangle}p_{i,\pm}(\varepsilon) \text{ if } j = i \pm 1, k < i < r, \\
{}_{\langle k,r\rangle}p_{r,\pm}(\varepsilon) = {}_{\langle k-1,r\rangle}p_{r,\pm}(\varepsilon) \text{ if } j = r - \frac{1\mp1}{2}, i = r, \\
0 \qquad\qquad \text{otherwise},
\end{cases} \tag{5.14}
$$

and

$$
{}_{\langle k,r\rangle}e_{ij}(\varepsilon) = \begin{cases}
{}_{\langle k,r\rangle}e_{k,+}(\varepsilon) = {}_{\langle k-1,r\rangle}e_{k,+}(\varepsilon) \\
\qquad \text{if } j = k+1, i = k, \\
{}_{\langle k,r\rangle}e_{k,-}(\varepsilon) = {}_{\langle k-1,r\rangle}e_{k,-}(\varepsilon) \\
\qquad + {}_{\langle k-1,r\rangle}e_{k-1}(\varepsilon) \cdot \frac{{}_{\langle k-1,r\rangle}p_{k,-}(\varepsilon)}{{}_{\langle k-1,r\rangle}p_{k-1,+}(\varepsilon)} \\
\qquad \text{if } j = k, i = k, \\
{}_{\langle k,r\rangle}e_{i,\pm}(\varepsilon) = {}_{\langle k-1,r\rangle}e_{i,\pm}(\varepsilon) \\
\qquad \text{if } j = i \pm 1, k < i < r, \\
{}_{\langle k,r\rangle}e_{r,\pm}(\varepsilon) = {}_{\langle k-1,r\rangle}e_{r,\pm}(\varepsilon) \\
\qquad \text{if } j = r - \frac{1\mp1}{2}, i = r, \\
0 \qquad \text{otherwise},
\end{cases} \tag{5.15}
$$

where

$$
{}_{\langle k,r\rangle}e_i(\varepsilon) = {}_{\langle k,r\rangle}e_{i,+}(\varepsilon) + {}_{\langle k,r\rangle}e_{i,-}(\varepsilon). \tag{5.16}
$$

In analogous way, transition probabilities ${}_{\langle k,r\rangle}p_{ij}(\varepsilon)$ and expectations of transition times ${}_{\langle k,r\rangle}e_{ij}(\varepsilon)$, for the reduced embedded semi-Markov process ${}_{\langle k,r\rangle}\eta_\varepsilon(t)$, can be expressed via transition probabilities ${}_{\langle k,r+1\rangle}p_{ij}(\varepsilon)$ and expectations of transition times ${}_{\langle k,r+1\rangle}e_{ij}(\varepsilon)$, for the reduced embedded semi-Markov process ${}_{\langle k,r+1\rangle}\eta_\varepsilon(t)$, for $0 \le k \le r \le N-1$, and, for every $\varepsilon \in (0, \varepsilon_0]$,

5.1.4 Explicit Formulas for Expectations of Return Times and Stationary Probabilities for Birth-Death-Type Semi-Markov Processes

As was mentioned above, the process ${}_{\langle 0,N\rangle}\eta_\varepsilon(t) = \eta_\varepsilon(t)$. Also, process ${}_{\langle 1,N\rangle}\eta_\varepsilon(t) = {}_0\eta_\varepsilon(t)$ and process ${}_{\langle 0,N-1\rangle}\eta_\varepsilon(t) = {}_N\eta_\varepsilon(t)$.

Thus, relations (5.14) and (5.15) reduce, respectively, to relations (5.8) and (5.9), when computing, respectively, the transition probabilities ${}_{\langle 1,N\rangle}p_{k,\pm}(\varepsilon)$ and expectations ${}_{\langle 1,N\rangle}e_{k,\pm}(\varepsilon)$, etc.

By iterating recurrent formulas (5.14)–(5.15) and their "right-hand" analogs, we get the following explicit formulas for transition probabilities ${}_{\langle k,r\rangle}p_{ij}(\varepsilon)$ and expectations of transition times ${}_{\langle k,r\rangle}e_{ij}(\varepsilon)$ for the reduced embedded semi-Markov process ${}_{\langle k,r\rangle}\eta_\varepsilon(t)$ expressed in terms of the transition characteristic for the initial semi-Markov process $\eta_\varepsilon(t)$, for $0 \leq k \leq r \leq N$, and, for every $\varepsilon \in (0, \varepsilon_0]$,

$$
{}_{\langle k,r\rangle}p_{ij}(\varepsilon) = \begin{cases} {}_{\langle k,r\rangle}p_{k,\pm}(\varepsilon) = p_{k,\pm}(\varepsilon) \text{ if } j = k + \frac{1\pm 1}{2}, i = k, \\ {}_{\langle k,r\rangle}p_{i,\pm}(\varepsilon) = p_{i,\pm}(\varepsilon) \text{ if } j = i \pm 1, k < i < r, \\ {}_{\langle k,r\rangle}p_{r,+}(\varepsilon) = p_{r,\pm}(\varepsilon) \text{ if } j = r - \frac{1\mp 1}{2}, i = r, \\ 0 \qquad\qquad \text{otherwise,} \end{cases} \tag{5.17}
$$

and

$$
{}_{\langle k,r\rangle}e_{ij}(\varepsilon) = \begin{cases} {}_{\langle k,r\rangle}e_{k,+}(\varepsilon) = e_{k,+}(\varepsilon) \\ \qquad \text{if } j = k+1, i = k, \\ {}_{\langle k,r\rangle}e_{k,-}(\varepsilon) = e_{k,-}(\varepsilon) + e_{k-1}(\varepsilon) \cdot \frac{p_{k,-}(\varepsilon)}{p_{k-1,+}(\varepsilon)} \\ \qquad + \cdots + e_0(\varepsilon) \cdot \frac{p_{1,-}(\varepsilon) \cdots p_{k,-}(\varepsilon)}{p_{0,+}(\varepsilon) \cdots p_{k-1,+}(\varepsilon)} \\ \qquad \text{if } j = k, i = k, \\ {}_{\langle k,r\rangle}e_{i,+}(\varepsilon) = e_{i,\pm}(\varepsilon) \\ \qquad \text{if } j = i \pm 1, k < i < r, \\ {}_{\langle k,r\rangle}e_{r,+}(\varepsilon) = e_{r,+}(\varepsilon) + e_{r+1}(\varepsilon) \cdot \frac{p_{r,+}(\varepsilon)}{p_{r+1,-}(\varepsilon)} \\ \qquad + \cdots + e_N(\varepsilon) \cdot \frac{p_{N-1,+}(\varepsilon) \cdots p_{r,+}(\varepsilon)}{p_{N,-}(\varepsilon) \cdots p_{r+1,-}(\varepsilon)} \\ \qquad \text{if } j = r, i = r, \\ {}_{\langle k,r\rangle}e_{r,-}(\varepsilon) = e_{r,-}(\varepsilon) \\ \qquad \text{if } j = r-1, i = r, \\ 0 \qquad \text{otherwise.} \end{cases} \tag{5.18}
$$

Let us denote by ${}_{\langle k,r\rangle}\tau_{\varepsilon,j}$ the hitting time for the state $j \in {}_{\langle k,r\rangle}\mathbb{X}$ for the reduced semi-Markov process ${}_{\langle k,r\rangle}\eta_\varepsilon(t)$.

By Theorem 3.1, $\mathsf{E}_i\tau_{\varepsilon,j} = \mathsf{E}_i\, {}_{\langle k,r\rangle}\tau_{\varepsilon,j}$, for $i, j \in {}_{\langle k,r\rangle}\mathbb{X}$ and $\varepsilon \in (0, \varepsilon_0]$,

Let us now choose $k = r = i \in \mathbb{X}$. In this case, the reduced phase space ${}_{\langle i,i\rangle}\mathbb{X} = \{i\}$ is a one-state set. Process ${}_{\langle i,i\rangle}\eta_\varepsilon(t)$ returns to the state i after every jump. This implies that, for every $\varepsilon \in (0, \varepsilon_0]$,

$$
E_{ii}(\varepsilon) = \mathsf{E}_i\tau_{\varepsilon,i} = \mathsf{E}_i\, {}_{\langle i,i\rangle}\tau_{\varepsilon,i} = {}_{\langle i,i\rangle}e_i(\varepsilon). \tag{5.19}
$$

The following formula takes place, for every $i \in \mathbb{X}$, and, for every $\varepsilon \in (0, \varepsilon_0]$,

$$\begin{aligned} E_{ii}(\varepsilon) &= e_i(\varepsilon) + e_{i-1}(\varepsilon)\frac{p_{i,-}(\varepsilon)}{p_{i-1,+}(\varepsilon)} + e_{i-2}(\varepsilon)\frac{p_{i-1,-}(\varepsilon)p_{i,-}(\varepsilon)}{p_{i-2,+}(\varepsilon)p_{i-1,+}(\varepsilon)} \\ &\quad + \cdots + e_0(\varepsilon)\frac{p_{1,-}(\varepsilon)p_{2,-}(\varepsilon)\cdots p_{i,-}(\varepsilon)}{p_{0,+}(\varepsilon)p_{1,+}(\varepsilon)\cdots p_{i-1,+}(\varepsilon)} \\ &\quad + e_{i+1}(\varepsilon)\frac{p_{i,+}(\varepsilon)}{p_{i+1,-}(\varepsilon)} + e_{i+2}(\varepsilon)\frac{p_{i+1,+}(\varepsilon)p_{i,+}(\varepsilon)}{p_{i+2,-}(\varepsilon)p_{i+1,-}(\varepsilon)} \\ &\quad + \cdots + e_N(\varepsilon)\frac{p_{N-1,+}(\varepsilon)p_{N-2,+}(\varepsilon)\cdots p_{i,+}(\varepsilon)}{p_{N,-}(\varepsilon)p_{N-1,-}(\varepsilon)\cdots p_{i+1,-}(\varepsilon)} \\ &= e_i(\varepsilon)U_{i,i}(\varepsilon) + e_{i-1}(\varepsilon)U_{i,i-1}(\varepsilon) + \cdots + e_0(\varepsilon)U_{1,0}(\varepsilon) \\ &\quad + e_{i+1}(\varepsilon)U_{i,i+1}(\varepsilon) + \cdots + e_N(\varepsilon)U_{i,N}(\varepsilon) \\ &= \sum_{j=0}^{N} e_j(\varepsilon)U_{i,j}(\varepsilon). \end{aligned} \tag{5.20}$$

Here and henceforth, we use the following useful notations:

$$U_{i,j}(\varepsilon) = \begin{cases} \frac{V_{j+1,i,-}(\varepsilon)}{V_{j,i-1,+}(\varepsilon)} & \text{for } 0 \le j < i \le N, \\ 1 & \text{for } 0 \le j = i \le N, \\ \frac{V_{i,j-1,+}(\varepsilon)}{V_{i+1,j,-}(\varepsilon)} & \text{for } 0 \le i < j \le N, \end{cases} \tag{5.21}$$

where

$$V_{i,j,\pm}(\varepsilon) = \prod_{k=i}^{j} p_{k,\pm}(\varepsilon), \text{ for } 0 \le i \le j \le N. \tag{5.22}$$

This is useful to note that quantity $U_{i,j}(\varepsilon)$ is the expectation of the number of visits of state j by the embedded Markov chain $\eta_{\varepsilon,n}$, before its first return to state i, i.e., for every $i, j \in \mathbb{X}$,

$$U_{i,j}(\varepsilon) = \mathsf{E}_i \sum_{n=1}^{\nu_{\varepsilon,i}} I(\eta_{\varepsilon,n-1} = j). \tag{5.23}$$

In particular,

$$\begin{aligned} E_{00}(\varepsilon) &= e_0(\varepsilon) + e_1(\varepsilon)\frac{p_{0,+}(\varepsilon)}{p_{1,-}(\varepsilon)} + e_2(\varepsilon)\frac{p_{1,+}(\varepsilon)p_{0,+}(\varepsilon)}{p_{2,-}(\varepsilon)p_{1,-}(\varepsilon)} \\ &\quad + \cdots + e_N(\varepsilon)\frac{p_{N-1,+}(\varepsilon)p_{N-2,+}(\varepsilon)\cdots p_{0,+}(\varepsilon)}{p_{N,-}(\varepsilon)p_{N-1,-}(\varepsilon)\cdots p_{1,-}(\varepsilon)} \\ &= e_0(\varepsilon)U_{0,0}(\varepsilon) + e_1(\varepsilon)U_{0,1}(\varepsilon) + \cdots + e_N(\varepsilon)U_{0,N}(\varepsilon). \end{aligned} \tag{5.24}$$

Formulas (3.18) and (5.20) yield, in an obvious way, explicit formulas for stationary and conditional quasi-stationary distributions for birth-death-type semi-Markov processes.

Lemma 5.1. *Let conditions* **J–L** *hold for semi-Markov processes $\eta_\varepsilon(t)$. Then, its stationary probabilities $\pi_i(\varepsilon), i \in \mathbb{X}$ are given by the following explicit formulas:*

$$\pi_i(\varepsilon) = \frac{e_i(\varepsilon)}{\sum_{j=0}^{N} e_j(\varepsilon) U_{i,j}(\varepsilon)}, \ i \in \mathbb{X}. \tag{5.25}$$

It should be noted that such formulas for stationary distributions of birth-death-type Markov chains are well known and can be found, for example, in Feller (1968). In context of our study, a special value has the presented above recurrent algorithm for getting such formulas, based on sequential reduction of the phase space for birth-death-type semi-Markov processes.

As far as explicit formulas for the conditional quasi-stationary probabilities $\pi_{\mathbb{U}_r,i}(\varepsilon)$, $i \in \mathbb{U}_r$ are concerned, they can be obtained by substituting the stationary probabilities $\pi_i(\varepsilon), i \in \mathbb{X}$, given by formulas (5.25), into formulas (5.5).

5.2 Asymptotic Expansions with Remainders Given the Standard Form for Perturbed Birth-Death-Type Semi-Markov Processes

In this section, we describe algorithms for construction of asymptotic expansions with remainders given in the standard form for stationary and conditional quasi-stationary distributions for perturbed birth-death-type semi-Markov processes.

5.2.1 Asymptotic Expansions with Remainders Given in the Standard Form for Expectations of Return Times and Related Functionals

The operational rules for Laurent asymptotic expansions formulated in Lemmas 2.1–2.4 can be applied to explicit expressions for the stationary and conditional quasi-stationary distributions for perturbed birth-death-type semi-Markov processes given in Section 5.1.

First, let us get asymptotic expansion for the products $V_{i,j,\pm}(\varepsilon)$ defined in relation (5.22).

Lemma 5.2. *Let conditions* **J–L** *and* $\mathbf{M}_L$ *hold. Then, function $V_{i,j,\pm}(\varepsilon)$ can, for every $0 \leq i \leq j \leq N$, be represented as the following pivotal Taylor asymptotic expansion:*

$$V_{i,j,\pm}(\varepsilon) = \sum_{r=l_{i,j,\pm}}^{L+l_{i,j,\pm}} v_{i,j,\pm}[r]\varepsilon^r + o_{v,i,j,\pm}(\varepsilon^{L+l_{i,j,\pm}}), \varepsilon \in (0, \varepsilon_0], \tag{5.26}$$

where:

(a) *Parameters $l_{i,j,\pm}$ are defined by the following relation:*

$$l_{i,j.\pm} = \sum_{i \leq r \leq j} l_{r.\pm}.$$

(b) *Coefficients $v_{i,j,\pm}[l_{i,j,\pm} + r], r = 0, \ldots, L$ are defined by the following relation:*

$$v_{i,j,\pm}[l_{i,j,\pm} + r] = \sum_{r_i + \cdots + r_j = r, 0 \leq r_k \leq L, i \leq k \leq j} \prod_{k=i}^{j} a_{k,\pm}[l_{k,\pm} + r_k].$$

(c) *Remainder $o_{v,i,j,\pm}(\varepsilon^{L+l_{i,j,\pm}})$ is defined by the following relation:*

$$o_{v,i,j,\pm}(\varepsilon^{L+l_{i,j,\pm}}) = \sum_{L+l_{i,j,\pm} < r_i + \cdots + r_j, l_{k,\pm} \leq r_k \leq L + l_{k,\pm}, i \leq k \leq j} \prod_{k=i}^{j} a_{k,\pm}[r_k] \varepsilon^{r_i + \cdots + r_j}$$
$$+ \sum_{1 \leq r \leq j-i+1} \sum_{i \leq m_1 < \cdots < m_r \leq j} \Big(\prod_{m \neq m_1, \ldots, m_r, i \leq m \leq j} \Big(\sum_{l_{m,\pm} \leq l \leq L + l_{m,\pm}} a_{m,\pm}[l] \varepsilon^l \Big)$$
$$\times \prod_{1 \leq k \leq r} o_{m_k,\pm}(\varepsilon^{L+l_{m_k,\pm}}) \Big).$$

(d) *If condition* $\mathbf{Q}_1$ *holds, then*: (1) $l_{i,j,\pm} = 0, v_{i,j,\pm}[0] > 0$, *for* $0 \leq i \leq j \leq N$.

(e) *If condition* $\mathbf{Q}_2$ *holds, then*: (1) $l_{0,j,-} = 0, v_{0,j,-}[0] > 0, l_{0,j,+} = 1, v_{0,j,+}[1] > 0$, *for* $0 \leq j \leq N$; (2) $l_{i,j,\pm} = 0, v_{i,j,\pm}[0] > 0$, *for* $0 < i \leq j \leq N$.

(f) *If condition* $\mathbf{Q}_3$ *holds, then*: (1) $l_{0,j,-} = 0, v_{0,j,-}[0] > 0, l_{0,j,+} = 1, v_{0,j,+}[1] > 0$, *for* $0 \leq j < N$; (2) $l_{0,N,\pm} = 1, v_{0,N,\pm}[1] > 0$; (3) $l_{i,j,\pm} = 0, v_{i,j,\pm}[0] > 0$, *for* $0 < i \leq j < N$; (4) $l_{i,N,-} = 1, v_{i,N,-}[1] > 0, l_{i,N,+} = 0, v_{i,N,+}[0] > 0$, *for* $0 < i \leq N$.

Proof. Asymptotic expansions for the probabilities $p_{i,\pm}(\varepsilon), 0 \leq i \leq N$ appearing in condition $\mathbf{M}_L$ let us apply proposition **(ii)** (the multiple multiplication rule) of Lemma 2.3 to products $V_{i,j,\pm}(\varepsilon), 0 \leq i \leq j \leq N$ given in relation (5.22).

Parameter $h_{V_{i,j,\pm}} = \sum_{i \leq r \leq j} l_{r.\pm} = l_{i,j,\pm}$ takes values 0 or 1 according to the rules described in propositions **(d)**–**(f)** of the lemma. Parameter $k_{V_{i,j,\pm}} = \min_{i \leq r \leq j}(L + l_{r,\pm} + \sum_{i \leq k \leq j, k \neq r} l_{k,\pm}) = L + l_{i,j,\pm}$.

Coefficients $v_{i,j,\pm}[r], r = 0, \ldots, L + l_{i,j,\pm}$ and remainders $o_{v,i,j,\pm}(\varepsilon^{L+l_{i,j,\pm}})$ are given, respectively, in propositions **(b)** and **(c)** of Lemma 5.2. By these propositions, coefficient $v_{i,j,\pm}[l_{i,j,\pm}] > 0$, i.e., the asymptotic expansion for function $V_{i,j,\pm}(\varepsilon)$ is pivotal.

Relations for parameter $l_{i,j,\pm}$ given in propositions **(d)**–**(f)** of Lemma 5.2 follow from the formula for these parameters given in proposition **(a)** of Lemma 5.2, and relations for parameters $l_{i,\pm}$ given in conditions $\mathbf{M}_L$ and $\mathbf{Q}_1$–$\mathbf{Q}_3$. □

Remark 5.1. Alternative recurrent formulas for the parameters, coefficients, and remainders of asymptotic expansions for products $V_{i,j,\pm}(\varepsilon)$ can be obtained by the use of the recurrent identities $V_{i,j,\pm}(\varepsilon) \equiv V_{i,j-1,\pm}(\varepsilon) \cdot p_{j,\pm}(\varepsilon), j = i, \ldots, N, 0 \leq i \leq N$, $V_{i,i-1,\pm}(\varepsilon) \equiv 1, 0 \leq i \leq N$, and the sequential application to them the operational rules for Laurent asymptotic expansions given in Lemma 2.2.

Second, let us get asymptotic expansions for the expectations $U_{i,j}(\varepsilon)$ defined in relation (5.21).

Lemma 5.3. *Let conditions* **J–L** *and* $\mathbf{M}_L$ *hold. Then, expectation* $U_{i,j}(\varepsilon)$ *can, for every* $0 \leq i, j \leq N$*, be represented as the following pivotal Laurent asymptotic expansion:*

$$U_{i,j}(\varepsilon) = \sum_{r=m_{i,j}}^{L+m_{i,j}} u_{i,j}[r]\varepsilon^r + o_{u,i,j}(\varepsilon^{L+m_{i,j}}), \varepsilon \in (0, \varepsilon_0], \tag{5.27}$$

where:

(a) *Parameter* $m_{i,j}$ *is defined by the following relation:*

$$m_{i,j} = \begin{cases} l_{j+1,i,-} - l_{j,i-1,+} & \text{for } 0 \leq j < i \leq N, \\ 0 & \text{for } 0 \leq j = i \leq N, \\ l_{i,j-1,+} - l_{i+1,j,-} & \text{for } 0 \leq i < j \leq N. \end{cases}$$

(b) *Coefficients* $u_{i,j}[m_{i,j}+r], r = 0, \ldots, L$ *are defined by the following relation:*

$$u_{i,j}[m_{i,j}+r] = \begin{cases} v_{j,i-1,+}[l_{j,i-1,+}]^{-1}(v_{j+1,i,-}[l_{j+1,i,-}+r] & \\ \quad - \sum_{1 \leq k \leq r} v_{j,i-1,+}[l_{j,i-1,+}+k]u_{i,j}[m_{i,j}+r-k]) & \text{for } 0 \leq j < i \leq N, \\ \mathrm{I}(r=0) & \text{for } 0 \leq j = i \leq N, \\ v_{i+1,j,-}[l_{i+1,j,-}]^{-1}(v_{j,i-1,+}[l_{j,i-1,+}+r] & \\ \quad - \sum_{1 \leq k \leq r} v_{i+1,j,-}[l_{i+1,j,-}+k]u_{i,j}[m_{i,j}+r-k]) & \text{for } 0 \leq i < j \leq N. \end{cases}$$

(c) *Remainder* $o_{u,i,j}(\varepsilon^{L+m_{i,j}})$ *is defined by the following relation:*

$$o_{u,i,j}(\varepsilon^{L+m_{i,j}})$$

$$= \begin{cases} \dfrac{o_{v,j+1,i,-}(\varepsilon^{L+l_{j+1,i,-}})}{\sum_{l_{j,i-1,+} \leq r \leq L+l_{j,i-1,+}} v_{j,i-1,+}[r] + o_{v,j,i-1,+}(\varepsilon^{d+l_{j,i-1,+}})} & \\ \quad - \dfrac{\sum_{L+l_{j+1,i,-} < r+k, l_{j,i-1,+} \leq r \leq L+l_{j,i-1,+}, m_{i,j} \leq k \leq L+m_{i,j}} v_{j,i-1,+}[r]u_{i,j}[k]\varepsilon^{r+k}}{\sum_{l_{j,i-1,+} \leq r \leq L+l_{j,i-1,+}} v_{j,i-1,+}[r] + o_{v,j,i-1,+}(\varepsilon^{L+l_{j,i-1,+}})} & \\ \quad - \dfrac{\sum_{m_{i,j} \leq k \leq L+m_{i,j}} u_{i,j}[k]\varepsilon^k o_{v,j,i-1,+}(\varepsilon^{L+l_{j,i-1,+}})}{\sum_{l_{j,i-1,+} \leq r \leq L+l_{j,i-1,+}} v_{j,i-1,+}[r] + o_{v,j,i-1,+}(\varepsilon^{L+l_{j,i-1,+}})} & \text{for } 0 \leq j < i \leq N, \\ 0 & \text{for } 0 \leq j = i \leq N, \\ \dfrac{o_{v,i,j-1,+}(\varepsilon^{L+l_{i,j-1,+}})}{\sum_{l_{i+1,j,-} \leq r \leq L+l_{i+1,j,-}} v_{i+1,j,-}[r] + o_{v,i+1,j,-}(\varepsilon^{L+l_{i+1,j,-}})} & \\ \quad - \dfrac{\sum_{L+l_{i,j-1,+} < r+k, l_{i+1,j,-} \leq r \leq L+l_{i+1,j,-}, m_{i,j} \leq k \leq L+m_{i,j}} v_{i+1,j,-}[r]u_{i,j}[k]\varepsilon^{r+k}}{\sum_{l_{i+1,j,-} \leq r \leq L+l_{i+1,j,-}} v_{i+1,j,-}[r] + o_{v,i+1,j,-}(\varepsilon^{L+l_{i+1,j,-}})} & \\ \quad - \dfrac{\sum_{m_{i,j} \leq k \leq L+m_{i,j}} u_{i,j}[k]\varepsilon^k o_{v,i+1,j,-}(\varepsilon^{L+l_{i+1,j,-}})}{\sum_{l_{i+1,j,-} \leq r \leq L+l_{i+1,j,-}} v_{i+1,j,-}[r] + o_{v,i+1,j,-}(\varepsilon^{L+l_{i+1,j,-}})} & \text{for } 0 \leq i < j \leq N. \end{cases}$$

(d) *If condition* $\mathbf{Q_1}$ *holds, then*: (1) $m_{i,j} = 0, u_{i,j}[0] > 0$, *for* $0 \leq i, j \leq N$.

(e) *If condition* $\mathbf{Q_2}$ *holds, then*: (1) $m_{0,0} = 0, u_{0,0}[0] > 0$; (2) $m_{0,j} = 1, u_{0,j}[1] > 0$, *for* $0 < j \leq N$; (3) $m_{i,0} = -1, u_{i,0}[-1] > 0$, *for* $0 < i \leq N$; (4) $m_{i,j} = 0, u_{i,j}[0] > 0$, *for* $0 < i, j \leq N$.

(f) *If condition* $\mathbf{Q_3}$ *holds, then*: (1) $m_{0,0} = 0, u_{0,0}[0] > 0$; (2) $m_{0,j} = 1, u_{0,j}[1] > 0$, *for* $0 < j < N$; (3) $m_{0,N} = 0, u_{0,N}[0] > 0$; (4) $m_{i,0} = -1, u_{i,0}[-1] > 0$, *for* $0 < i < N$; (5) $m_{i,j} = 0, u_{i,j}[0] > 0$, *for* $0 < i, j < N$; (6) $m_{i,N} = -1, u_{i,N}[-1] > 0$, *for* $0 < i < N$; (7) $m_{N,0} = 0, u_{N,0}[0] > 0$; (8) $m_{N,j} = 1, u_{N,j}[1] > 0$, *for* $0 < j < N$; (9) $m_{N,N} = 0, u_{N,N}[0] > 0$.

Proof. Asymptotic expansions for functions $V_{i,j,\pm}(\varepsilon), 0 \leq i \leq j \leq N$, given in Lemma 5.2, let us apply proposition **(v)** (the division rule) of Lemma 2.2 to quotients $U_{i,j}(\varepsilon), 0 \leq i, j \leq N$ defined in relation (5.21).

Let us check formulas for parameters $m_{i,j}$ given in proposition **(a)** of the lemma. For example, let $0 \leq j < i \leq N$. In this case, $h_{U_{i,j}} = h_{V_{j+1,i,-}} - h_{V_{j,i-1,+}} = l_{j+1,i,-} - l_{j,i-1,+} = m_{i,j}$ and $k_{U_{i,j}} = (k_{V_{j+1,i,-}} - h_{V_{j,i-1,+}}) \wedge (k_{V_{j,i-1,+}} - 2h_{V_{j,i-1,+}} + h_{V_{j+1,i,-}}) = (L + l_{V_{j+1,i,-}} - l_{V_{j,i-1,+}}) \wedge (L + l_{j,i-1,+} - 2l_{j,i-1,+} + l_{j+1,i,-}) = L + l_{j+1,i,-} - l_{j,i-1,+} = L + m_{i,j}$. The check for the case, $0 \leq i < j \leq N$, is analogous. The case $0 \leq i = j \leq N$ is trivial.

Coefficients $u_{i,j}[m_{i,j} + r], r = 0, \ldots, L$ and remainders $o_{u,i,j,\pm}(\varepsilon^{L+m_{i,j}})$ are given, respectively, in propositions **(b)** and **(c)** of the lemma. By this proposition, coefficient $u_{i,j}[m_{i,j}] > 0$, i.e., the asymptotic expansion for expectation $U_{i,j}(\varepsilon)$ is pivotal.

Relations for parameter $m_{i,j}$ given in propositions **(d)**–**(f)** of Lemma 5.3 follow from the formula given in proposition **(a)** of this lemma and relations for parameters $l_{i,j,\pm}$ given in propositions **(d)**–**(f)** of Lemma 5.2. □

The following lemma gives asymptotic expansions for expectations $e_i(\varepsilon)$ defined in relation (5.10).

Lemma 5.4. *Let conditions* **J**–**L** *and* $\mathbf{N}_L$ *hold. Then, expectation* $e_i(\varepsilon)$ *can, for every* $0 \leq i \leq N$, *be represented as the following pivotal Taylor asymptotic expansion:*

$$e_i(\varepsilon) = \sum_{r=0}^{L} b_i[r]\varepsilon^r + \dot{o}_i(\varepsilon^L), \varepsilon \in (0, \varepsilon_0], \tag{5.28}$$

where:

(a) *Coefficients* $b_i[r], r = 0, \ldots, L$ *are defined by the following relation:*

$$b_i[r] = b_{i,-}[r] + b_{i,+}[r].$$

where $b_{i,\pm}[0] = 0$, *if* $l_{i,\pm} = 1$.

(b) *Remainder* $\dot{o}_i(\varepsilon^L)$ *is defined by the following relation:*

$$\dot{o}_i(\varepsilon^L) = \begin{cases} \dot{o}_{0,-}(\varepsilon^L) + b_{0,-}[L+1]\varepsilon^{L+1} + \dot{o}_{0,-}(\varepsilon^{L+1}) & \text{if } i = 0, \\ \dot{o}_{i,-}(\varepsilon^L) + \dot{o}_{i,+}(\varepsilon^L) & \text{if } 0 < i < N, \\ b_{N,-}[L+1]\varepsilon^{L+1} + \dot{o}_{N,-}(\varepsilon^{L+1}) + \dot{o}_{N,+}(\varepsilon^L) & \text{if } i = N. \end{cases}$$

(c) *If conditions* **P** *and* $\mathbf{Q}_1$ *hold, then*: (1) $l_{i,\pm} = 0, b_{i,\pm}[0] > 0$, *and, thus,* $b_i[0] > 0$, *for* $0 \leq i \leq N$.

(d) *If conditions* **P** *and* $\mathbf{Q}_2$ *hold, then*: (1) $l_{0,-} = 0, b_{0,-}[0] > 0, l_{0,+} = 1, b_{0,+}[0] = 0, b_{0,+}[1] > 0$ *and, thus,* $b_0[0] > 0$; (2) $l_{i,\pm} = 0, b_{i,\pm}[0] > 0$, *and, thus,* $b_i[0] > 0$, *for* $0 < i \leq N$.

(e) *If conditions* **P** *and* $\mathbf{Q}_3$ *hold, then*: (1) $l_{0,-} = 0, b_{0,-}[0] > 0, l_{0,+} = 1, b_{0,+}[0] = 0, b_{0,+}[1] > 0$, *and, thus,* $b_0[0] > 0$; (2) $l_{i,\pm} = 0, b_{i,\pm}[0] > 0$, *and, thus,* $b_i[0] > 0$, *for* $0 < i < N$; (3) $l_{N,-} = 1, b_{N,-}[0] = 0, b_{N,-}[1] > 0, l_{N,+} = 0, b_{N,+}[0] > 0$, *and, thus,* $b_N[0] > 0$.

Proof. It is done by application of proposition **(ii)** (the summation rule) of Lemma 2.2 to sums $e_i(\varepsilon) = e_{i,-}(\varepsilon) + e_{i,+}(\varepsilon), 0 \leq i \leq N$.

In this case, parameters $h_{e_i} = l_{i,-} \wedge l_{i,+} = 0$ and $k_{e_i} = (L + l_{i,-}) \wedge (L + l_{i,+}) = L$.

Coefficients $b_i[r], r = 0, \ldots, L$ and remainder $\dot{o}_i(\varepsilon^L)$ are defined, respectively, by relations given in propositions **(b)** and **(c)** of the lemma. As follows from condition $\mathbf{N}_L$, in this case coefficient $b_i[0] > 0$, i.e., the asymptotic expansion for expectation $e_i(\varepsilon)$ is pivotal.

Relations for parameters $l_{i,\pm}$ given in propositions **(c)**–**(e)** of Lemma 5.4 follow from the formula given in proposition **(a)** of this lemma and relations for parameters $l_{i,\pm}$ given in condition $\mathbf{N}_L$, **P**, and $\mathbf{Q}_1$–$\mathbf{Q}_3$. □

Remark 5.2. In Lemma 5.4, condition $\mathbf{M}_L$ is not required. In this case, parameters $a_i[0]$ should be replaced by parameters $b_i[0]$ in conditions $\mathbf{Q}_1$, $\mathbf{Q}_2$, and $\mathbf{Q}_3$. This replacement is formally realized by condition **P**.

Let us consider functions,

$$W_{i,j}(\varepsilon) = e_j(\varepsilon) U_{i,j}(\varepsilon), \text{ for } i, j \in \mathbb{X}. \tag{5.29}$$

This is useful to note that quantity $W_{i,j}(\varepsilon)$ is the time spent by process $\eta_\varepsilon(t)$ in state j, before its first return to state i, i.e., for every $i, j \in \mathbb{X}$,

$$W_{i,j}(\varepsilon) = \mathsf{E}_i \sum_{n=1}^{\nu_{\varepsilon,i}} I(\eta_{\varepsilon,n-1} = j) \kappa_{\varepsilon,n}. \tag{5.30}$$

Lemma 5.5. *Let conditions* **J**–**L**, $\mathbf{M}_L$, $\mathbf{N}_L$, *and* **P** *hold. Then, expectation* $W_{i,j}(\varepsilon)$ *can, for every* $0 \leq i, j \leq N$, *be represented as the following pivotal Laurent asymptotic expansion:*

$$W_{i,j}(\varepsilon) = \sum_{r=m_{i,j}}^{L+m_{i,j}} w_{i,j}[r]\varepsilon^r + \dot{o}_{w,i}(\varepsilon^{L+m_{i,j}}), \varepsilon \in (0, \varepsilon_0], \tag{5.31}$$

where:

(a) *Coefficients* $w_{i,j}[m_{i,j} + r], r = 0, \ldots, L$ *are defined by the following relation:*

$$w_{i,j}[m_{i,j} + r] = \sum_{0 \leq k \leq r} e_j[k] u_{i,j}[m_{i,j} + r - k].$$

(b) *Remainder* $o_{w,i,j}(\varepsilon^{L+m_{i,j}})$ *is defined by the following relation:*

$$\begin{aligned} o_{w,i,j}(\varepsilon^{L+m_{i,j}}) = & \sum_{L+m_{i,j}<r+k, 0\leq r\leq L, m_{i,j}\leq k\leq L+m_{i,j}} b_j[r]u_{i,j}[k]\varepsilon^{r+k} \\ & + \sum_{0\leq r\leq L} b_j[r]\varepsilon^r o_{u,i,j}(\varepsilon^{L+m_{i,j}}) + \sum_{m_{i,j}\leq k\leq L+m_{i,j}} u_{i,j}[k]\varepsilon^k \dot{o}_j(\varepsilon^L) \\ & + \dot{o}_j(\varepsilon^L) o_{u,i,j}(\varepsilon^{L+m_{i,j}}). \end{aligned}$$

(c) *If condition* $\mathbf{Q_1}$ *holds, then*: (1) $m_{i,j} = 0, w_{i,j}[0] > 0$, *for* $0 \leq i, j \leq N$.

(d) *If condition* $\mathbf{Q_2}$ *holds, then*: (1) $m_{0,0} = 0, w_{0,0}[0] > 0$; (2) $m_{0,j} = 1, w_{0,j}[1] > 0$, *for* $0 < j \leq N$; (3) $m_{i,0} = -1, w_{i,0}[-1] > 0$, *for* $0 < i \leq N$; (4) $m_{i,j} = 0, w_{i,j}[0] > 0$, *for* $0 < i, j \leq N$

(e) *If condition* $\mathbf{Q_3}$ *holds, then*: (1) $m_{0,0} = 0, w_{0,0}[0] > 0$; (2) $m_{0,j} = 1, w_{0,j}[1] > 0$, *for* $0 < j < N$; (3) $m_{0,N} = 0, w_{0,N}[0] > 0$; (4) $m_{i,0} = -1, w_{i,0}[-1] > 0$, *for* $0 < i < N$; (5) $m_{i,j} = 0, w_{i,j}[0] > 0$, *for* $0 < i, j < N$; (6) $m_{i,N} = -1, w_{i,N}[-1] > 0$, *for* $0 < i < N$; (7) $m_{N,0} = 0, w_{N,0}[0] > 0$; (8) $m_{N,j} = 1, w_{N,j}[1] > 0$, *for* $0 < j < N$; (9) $m_{N,N} = 0, w_{N,N}[0] > 0$.

Proof. Asymptotic expansions for functions $U_{i,j}(\varepsilon), 0 \leq i, j \leq N$ and $e_j(\varepsilon), 0 \leq j \leq N$, given, respectively, in Lemmas 5.3 and 5.4, let us apply proposition **(iii)** (the multiplication rule) of Lemma 2.2 to products $W_{i,j}(\varepsilon), 0 \leq i, j \leq N$ defined in relation (5.29).

In this case, $h_{W_{i,j}} = h_{e_{ij}} + h_{U_{i,j}} = 0 + m_{i,j} = m_{i,j}, k_{W_{i,j}} = (k_{e_{ij}} + h_{U_{i,j}}) \wedge (k_{U_{ij}} + h_{e_{i,j}}) = (L + m_{i,j}) \wedge (L + m_{i,j} + 0) = L + m_{i,j}$.

Coefficients $w_{i,j}[m_{i,j} + r], r = 0, \ldots, L$ and remainders $o_{w,i,j,\pm}(\varepsilon^{L+m_{i,j}})$ are given, respectively, in propositions **(a)** and **(b)** of the lemma. According to proposition **(iii)** of Lemma 2.2 the asymptotic expansion for function $W_{i,j}(\varepsilon)$ is pivotal.

Relations for parameter $m_{i,j}$ and given in propositions **(c)**–**(e)** of Lemma 5.5 follow from relations for parameters $m_{i,j}$ given in propositions **(d)**–**(f)** of Lemma 5.3, and the pivotal character of asymptotic expansion for function $e_j(\varepsilon)$. □

The following lemma gives Laurent asymptotic expansions for expectations of return times,

$$E_{ii}(\varepsilon) = W_{i,0}(\varepsilon) + \cdots + W_{i,N}(\varepsilon), i \in \mathbb{X}. \tag{5.32}$$

Lemma 5.6. *Let conditions* **J–L**, $\mathbf{M}_L$, $\mathbf{N}_L$, *and* **P** *hold. Then, expectation* $E_{ii}(\varepsilon)$ *can, for every* $0 \leq i \leq N$, *be represented as the following pivotal Laurent asymptotic expansion:*

$$E_{ii}(\varepsilon) = \sum_{r=M_{ii}}^{L+M_{ii}} B_{ii}[r]\varepsilon^r + \dot{o}_{ii}(\varepsilon^{M_{ii}}), \ \varepsilon \in (0, \varepsilon_0], \tag{5.33}$$

where:

(a) *Parameter* M_{ii} *is defined by the following relation:*

$$M_{ii} = \min_{j\in\mathbb{X}} m_{i,j}.$$

(b) *Coefficients* $B_{ii}[M_{ii}+r], r=0,\ldots,L$ *are defined by the following relation:*

$$B_{ii}[M_{ii}+r] = \sum_{j\in\mathbb{X}} w_{i,j}[M_{ii}+r],$$

where $w_{i,j}[M_{ii}+r]=0$*, for* $0\le r<m_{i,j}-M_{ii}, j\in\mathbb{X}$.

(c) *Remainder* $\dot{o}_{ii}(\varepsilon^{L+M_{ii}})$ *is defined by the following relation:*

$$\dot{o}_{ii}(\varepsilon^{L+M_{ii}}) = \sum_{j\in\mathbb{X}}\Big(\sum_{L+M_{ii}<r\le L+m_{i,j}} w_{i,j}[r]\varepsilon^r + o_{w,i,j}(\varepsilon^{L+m_{i,j}})\Big).$$

(d) *If condition* $\mathbf{Q_1}$ *holds, then*: (1) $M_{ii}=0, B_{ii}[0]>0$*, for* $0\le i\le N$.

(e) *If condition* $\mathbf{Q_2}$ *holds, then*: (1) $M_{00}=0, B_{00}[0]>0$; (2) $M_{ii}=-1, B_{ii}[-1]>0$*, for* $0<i\le N$.

(f) *If condition* $\mathbf{Q_3}$ *holds, then*: (1) $M_{00}=0, B_{00}[0]>0$; (2) $M_{ii}=-1, B_{ii}[-1]>0$*, for* $0<i<N$; (3) $M_{NN}=0, B_{NN}[0]>0$.

Proof. Asymptotic expansions for functions $W_{i,j}(\varepsilon), 0\le i,j\le N$, given in Lemma 5.5, let us apply proposition **(i)** (the multiple summation rule) of Lemma 2.3 to sums $E_{ii}(\varepsilon), 0\le i\le N$ defined in relation (5.32).

In this case, $h_{E_{ii}}=\min_{0\le j\le N} m_{i,j}=M_{ii}, k_{E_{ii}}=\min_{0\le j\le N}(L+m_{i,j})=L+M_{ii}$.

Coefficients $B_{ii}[M_{i,j}+r], r=0,\ldots,L$ and remainder $\dot{o}_{ii}(\varepsilon^{L+M_{ii}})$ are given, respectively, in propositions **(b)** and **(c)** of the lemma. According to proposition **(i)** of Lemma 2.3 the asymptotic expansion for function $E_{ii}(\varepsilon)$ is pivotal.

Relations for parameter M_{ii} given in propositions **(d)**–**(f)** of Lemma 5.6 follow from the formula given in proposition **(a)** of this lemma and relations for parameters $m_{i,j}$ given in propositions **(c)**–**(e)** of Lemma 5.5. □

Remark 5.3. Alternative recurrent formulas for the parameters, coefficients, and remainders of asymptotic expansions for expectations $E_{ii}(\varepsilon)=W_{i,0}(\varepsilon)+\cdots+W_{i,N}(\varepsilon)$ can be obtained by the use of the recurrent identities $E_{ii,n}(\varepsilon)\equiv W_{i,0}(\varepsilon)+\cdots+W_{i,n}(\varepsilon)=E_{ii,n-1}(\varepsilon)+W_{i,n}(\varepsilon), n=0,\ldots N, 0\le i\le N$, $E_{ii,-1}(\varepsilon)\equiv 0, 0\le i\le N$, and the sequential application to them the operational rules for Laurent asymptotic expansions given in Lemma 2.2.

5.2.2 Asymptotic Expansions with Remainders Given in the Standard Form for Stationary and Conditional Quasi-Stationary Distributions

Now, we are prepared to give Taylor asymptotic expansions for stationary probabilities of nonlinearly perturbed birth-death-type semi-Markov processes.

Theorem 5.1. *Let conditions* **J–L**, $\mathbf{M}_L$, $\mathbf{N}_L$, *and* **P** *hold. Then, the stationary probability $\pi_i(\varepsilon)$ can, for every $0 \leq i \leq N$, be represented as the following pivotal Taylor asymptotic expansion:*

$$\pi_i(\varepsilon) = \sum_{r=n_i}^{L+n_i} c_i[r]\varepsilon^r + o_i(\varepsilon^{L+n_i}), \varepsilon \in (0, \varepsilon_0], \tag{5.34}$$

where:

(a) *Parameter n_i is defined by the following relation:*

$$n_i = -M_{ii}.$$

(b) *Coefficients $c_i[n_i+k], k = 0, \ldots, L$ are defined by the following recurrent relation:*

$$c_i[n_i+k] = B_{ii}[M_{ii}]^{-1}(b_i[k] - \sum_{1\leq l\leq r} B_{ii}[M_{ii}+l]c_i[n_i+k-l]).$$

(c) *Remainder $o_i(\varepsilon^{L+m_i})$ is defined by the following relation:*

$$o_i(\varepsilon^{L+n_i}) = \frac{\dot{o}_i(\varepsilon^L) - \sum_{L<r+k, M_{ii}\leq r\leq L+M_{ii}, n_i\leq k\leq L+n_i} B_{ii}[r]c_i[k]\varepsilon^{r+k}}{\sum_{M_{ii}\leq r\leq L+M_{ii}} B_{ii}[r]\varepsilon^r + \ddot{o}_{ii}(\varepsilon^{L+M_{ii}})}$$
$$- \frac{\sum_{n_i\leq k\leq L+n_i} c_i[k]\varepsilon^k \ddot{o}_{ii}(\varepsilon^{L+M_{ii}})}{\sum_{M_{ii}\leq r\leq L+M_{ii}} B_{ii}[r]\varepsilon^r + \ddot{o}_{ii}(\varepsilon^{L+M_{ii}})}.$$

(d) *If condition* $\mathbf{Q}_1$ *holds, then*: (1) $n_i = 0, c_i[0] > 0$, *for* $0 \leq i \leq N$; (2) $\sum_{0\leq i\leq N} c_i[0] = 1, \sum_{0\leq i\leq N} c_i[r] = 0, r = 1, \ldots, L$; (3) $\sum_{0\leq i\leq N} o_i(\varepsilon^L) = 0, \varepsilon \in (0, \varepsilon_0]$.

(e) *If condition* $\mathbf{Q}_2$ *holds, then*: (1) $n_0 = 0, c_0[0] > 0$; (2) $n_i = 1, c_i[1] > 0$, *for* $0 < i \leq N$; (3) $c_0[0] = 1, c_i[0] = 0, 0 < i \leq N, \sum_{0\leq i\leq N} c_i[r] = 0, r = 1, \ldots, L$; (4) $o_0(\varepsilon^L) + \sum_{0<i\leq N}(c_i[L+1] + o_i(\varepsilon^{L+1})) = 0, \varepsilon \in (0, \varepsilon_0]$.

(f) *If condition* $\mathbf{Q}_3$ *holds, then*: (1) $n_0 = 0, c_0[0] > 0$; (2) $n_i = 1, c_i[1] > 0$, *for* $0 < i < N$; (3) $n_N = 0, c_N[0] > 0$; (4) $c_0[0] + c_N[0] = 1, c_i[0] = 0, 0 < i < N, \sum_{0\leq i\leq N} c_i[r] = 0, r = 1, \ldots, L$; (5) $o_0(\varepsilon^L) + o_N(\varepsilon^L) + \sum_{0<i<N}(c_i[L+1] + o_i(\varepsilon^{L+1})) = 0, \varepsilon \in (0, \varepsilon_0]$.

Proof. The asymptotic expansions for expectations $e_i(\varepsilon), E_{ii}(\varepsilon), 0 \leq i \leq N$, given in Lemmas 5.4, and 5.6 let us apply proposition **(v)** (the division rule) of Lemma 2.2 to the stationary probabilities $\pi_i(\varepsilon) = \frac{e_i(\varepsilon)}{E_{ii}(\varepsilon)}, 0 \leq i \leq N$.

In this case, $h_{\pi_i} = h_{e_i} - h_{E_{ii}} = 0 - M_{ii} = n_i, k_{\pi_i} = (k_{e_i} - h_{E_{ii}}) \wedge (k_{E_{ii}} - 2h_{E_{ii}} + h_{e_i} = (L - M_{ii}) \wedge (L + M_{ii} - 2M_{ii} + 0) = L - M_{ii} = L + n_i$.

Coefficients $c_i[n_i + r], r = 0, \ldots, L$ and remainder $o_i(\varepsilon^{L+n_i})$ are given, respectively, in propositions **(b)** and **(c)** of the theorem. According to proposition **(v)** of Lemma 2.2, the asymptotic expansion for function $\pi_i(\varepsilon)$ is pivotal.

Relations for parameter n_i given in propositions **(d)**–**(f)** of Theorem 5.1 follow from the formula given in proposition **(a)** of this theorem, relations for parameter M_{ii} given in propositions **(d)**–**(f)** of Lemma 5.6, and the pivotal character of asymptotic expansion for function $e_i(\varepsilon)$ given Lemma 5.4.

Additional relations for coefficients $c_i[n_i+k], i \in \mathbb{X}, k = 0, \ldots, L$ and remainders $o_i(\varepsilon^{L+n_i}), i \in \mathbb{X}$, given in propositions **(d)**–**(f)** of Theorem 5.1 follow from the identity $\sum_{i \in \mathbb{X}} \pi_i(\varepsilon) = 1, \varepsilon \in (0, \varepsilon_0]$. □

Finally, let us give Taylor asymptotic expansions for the conditional quasi-stationary probabilities of nonlinearly perturbed birth-death-type semi-Markov processes defined in relation (5.5).

In order to do this, we, first, get asymptotic expansions for functions,

$$\Pi_{\mathbb{U}_r}(\varepsilon) = \sum_{i \in \mathbb{U}_r} \pi_i(\varepsilon). \tag{5.35}$$

Lemma 5.7. *Let conditions* **J**–**L**, $\mathbf{M}_L$, $\mathbf{N}_L$, *and* **P** *hold. Then, function* $\Pi_{\mathbb{U}_r}(\varepsilon)$ *can, for every* $r = 2,3$, *be represented as the following pivotal Taylor asymptotic expansion:*

$$\Pi_{\mathbb{U}_r}(\varepsilon) = \sum_{k=1}^{L+1} C_{\mathbb{U}_r}[k]\varepsilon^k + o_{\mathbb{U}_r}(\varepsilon^{L+1}), \ \varepsilon \in (0, \varepsilon_0], \tag{5.36}$$

where:

(a) *Coefficients* $C_{\mathbb{U}_r}[k], k = 1, \ldots, L+1$ *are defined by the following relation:*

$$C_{\mathbb{U}_r}[k] = \sum_{i \in \mathbb{U}_r} c_i[k].$$

(b) *Remainder* $o_{\mathbb{U}_r}(\varepsilon^{L+1})$ *is defined by the following relation:*

$$o_{\mathbb{U}_r}(\varepsilon^{L+1}) = \sum_{i \in \mathbb{U}_r} o_i(\varepsilon^{L+1}).$$

Proof. The asymptotic expansions for stationary probabilities $\pi_i(\varepsilon), i \in \mathbb{X}$, given in Theorem 5.1 let us apply proposition **(i)** (the multiple summation rule) of Lemma 2.3 to sums $\Pi_{\mathbb{U}_r}(\varepsilon)$ defined in relation (5.35).

Propositions **(e)** and **(f)** of Lemma 5.7 imply that all parameters $n_i = 1, i \in \mathbb{U}_r$, for $r = 2,3$. This implies that $h_{\Pi_{\mathbb{U}_r}} = \min_{i \in \mathbb{U}_r} n_i = 1, k_{\Pi_{\mathbb{U}_r}} = \min_{i \in \mathbb{U}_r}(L + n_i) = L+1$.

Coefficients $C_{\mathbb{U}_r}[k], k = 1, \ldots, L+1$ and remainder $o_{\mathbb{U}_r}(\varepsilon^{L+1})$ are given, respectively, in propositions **(a)** and **(b)** of the lemma. According to proposition **(i)** of Lemma 2.3, the asymptotic expansion for function $\Pi_{\mathbb{U}_r}(\varepsilon)$ is pivotal. □

Theorem 5.2. *Let conditions* **J**–**L**, $\mathbf{M}_L$, $\mathbf{N}_L$, *and* **P** *hold. Then, the conditional quasi-stationary probability* $\pi_{\mathbb{U}_r,i}(\varepsilon)$ *can, for every* $i \in \mathbb{U}_r$, $r = 2,3$, *be represented as the following pivotal Taylor asymptotic expansion:*

$$\pi_{\mathbb{U}_r,i}(\varepsilon) = \sum_{k=0}^{L} c_{\mathbb{U}_r,i}[k]\varepsilon^k + o_{\mathbb{U}_r,i}(\varepsilon^L), \varepsilon \in (0, \varepsilon_0], \tag{5.37}$$

where:

(a) *Coefficients* $c_{\mathbb{U}_r,i}[k], k = 0, \ldots, L$ *are defined by the following recurrent relation:*

$$c_{\mathbb{U}_r,i}[k] = C_{\mathbb{U}_r}[1]^{-1}(c_i[k+1] - \sum_{1\leq l\leq k} C_{\mathbb{U}_r}[l+1]c_{\mathbb{U}_r,i}[k-l+1]).$$

(b) *Remainder* $o_{\mathbb{U}_r,i}(\varepsilon^L)$ *is defined by the following relation:*

$$o_{\mathbb{U}_r,i}(\varepsilon^L) = \frac{o_i(\varepsilon^{L+1}) - \sum_{L<l+k,1\leq l\leq L+1,1\leq k\leq L+1} C_{\mathbb{U}_r}[l]c_{\mathbb{U}_r,i}[k]\varepsilon^{l+k}}{\sum_{1\leq l\leq L+1} C_{\mathbb{U}_r}[l]\varepsilon^l + o_{\mathbb{U}_r}(\varepsilon^{L+1})} - \frac{\sum_{1\leq k\leq L+1} c_{\mathbb{U}_r,i}[k]\varepsilon^k o_{\mathbb{U}_r}(\varepsilon^{L+1})}{\sum_{1\leq l\leq L+1} C_{\mathbb{U}_r}[l]\varepsilon^l + o_{\mathbb{U}_r}(\varepsilon^{L+1})}.$$

(c) (1) $c_{\mathbb{U}_r,i}[0] > 0$, *for* $i \in \mathbb{U}_r$; (2) $\sum_{i\in\mathbb{U}_r} c_{\mathbb{U}_r,i}[0] = 1$, $\sum_{i\in\mathbb{U}_r} c_i[l] = 0$, *for* $l = 1,\dots,L$; (3) $\sum_{i\in\mathbb{U}_r} o_{\mathbb{U}_r,i}(\varepsilon^L) = 0, \varepsilon \in (0,\varepsilon_0]$.

Proof. Asymptotic expansions for stationary probabilities $\pi_i(\varepsilon), i \in \mathbb{X}$, given in Teorem 5.1 let us apply proposition **(v)** (the division rule) of Lemma 2.2 to quotients $\pi_{\mathbb{U}_r,i}(\varepsilon) = \frac{\pi_i(\varepsilon)}{\Pi_{\mathbb{U}_r}(\varepsilon)}, i \in \mathbb{U}_r$, for $r = 2,3$.

Propositions **(e)** and **(f)** of Lemma 5.7 imply that all parameters $n_i = 1, i \in \mathbb{U}_r$, for $r = 2,3$. This implies that $h_{\pi_{\mathbb{U}_r,i}} = h_{\pi_i} - h_{\mathbb{U}_r} = 1 - 1 = 0$ and $k_{\pi_{\mathbb{U}_r,i}} = (k_{\pi_i} - h_{\mathbb{U}_r}) \wedge (h_{\mathbb{U}_r} - 2h_{\mathbb{U}_r} + h_{\pi_i}) = (L+1-1) \wedge (L+1-2\cdot 1+1) = L$.

Coefficients $c_{\mathbb{U}_r,i}[k], k = 0,\dots,L$ and remainder $o_{\mathbb{U}_r,i}(\varepsilon^L)$ are given, respectively, in propositions **(a)** and **(b)** of the theorem. According to proposition **(i)** of Lemma 2.3, the asymptotic expansion for function $\pi_{\mathbb{U}_r,i}(\varepsilon)$ is pivotal.

Additional relations for coefficients $c_{\mathbb{U}_r,i}[k], i \in \mathbb{U}_r, k = 0,\dots,L$ and remainders $o_{\mathbb{U}_r,i}(\varepsilon^L), i \in \mathbb{U}_r$, given in proposition **(c)** of Theorem 5.2 follow from the identity $\sum_{i\in\mathbb{U}_r} \pi_{\mathbb{U}_r,i}(\varepsilon) = 1, \varepsilon \in (0,\varepsilon_0]$. □

5.2.3 *Asymptotic Expansions with Remainders Given in the Standard Form and the Algorithm of Sequential Phase Space Reduction*

The operational rules for Laurent asymptotic expansions formulated in Lemmas 2.1–2.4 can be applied to the recurrent formulas (5.14)–(5.16) for transition probabilities and expectations of transition times for reduced perturbed birth-death semi-Markov processes.

The corresponding alternative recurrent algorithm would give the same Laurent asymptotic expansions for expectations $E_{ii}(\varepsilon), i \in \mathbb{X}$, as the algorithm presented in Lemmas 5.2–5.6. The only difference would be in forms of formulas for parameters, coefficients, and remainders in the corresponding expansions.

In the first case, these formulas would not such explicit as in the second. However, they would have a recurrent form better prepared for numerical computations and programming.

As far as asymptotic expansions for stationary and conditional quasi-stationary probabilities are concerned, they can be obtained using the algorithms described in Theorems 5.1 and 5.2, i.e., by application of the proposition **(ii)** (the division rule) of Lemma 2.2 to the corresponding quotients defining the stationary and conditional quasi-stationary probabilities.

5.3 Asymptotic Expansions with Explicit Upper Bounds for Remainders for Perturbed Birth-Death-Type Semi-Markov Processes

In this section, we describe algorithms for construction of asymptotic expansions, with explicit upper bounds for remainders, for stationary and conditional quasi-stationary distributions for perturbed birth-death-type semi-Markov processes.

5.3.1 Asymptotic Expansions with Explicit Upper Bounds for Remainders for Expectations of Return Times and Related Functionals

The operational rules for Laurent asymptotic expansions with explicit upper bounds for remainders formulated in Lemmas 2.5–2.7 can be applied to asymptotic expansions given in Lemmas 5.2–5.7 and Theorems 5.1–5.2, in order to get asymptotic expansions with explicit upper bounds for remainders for stationary and conditional quasi-stationary distributions for perturbed birth-death-type semi-Markov processes. We do this using in all cases simplified upper bounds for remainders given in Lemmas 2.5–2.7.

First, let us get asymptotic expansion with explicit upper bounds for remainders for the products $V_{i,j,\pm}(\varepsilon)$ defined in relation (5.22).

Lemma 5.8. *Let conditions* **J–L** *and* $\mathbf{M}'_L$ *hold. Then, for every* $0 \leq i \leq j \leq N$, *the pivotal* $(l_{i,j,\pm}, L + l_{i,j,\pm})$*-expansion for function* $V_{i,j,\pm}(\varepsilon)$ *given in Lemma 5.2 and Remark 5.1 is a pivotal* $(l_{i,j,\pm}, L + l_{i,j,\pm}, \delta_{v,i,j,\pm}, G_{v,i,j,\pm}, \varepsilon_{v,i,j,\pm})$*-expansion, with parameters* $\delta_{v,i,j,\pm}, G_{v,i,j,\pm}$ *and* $\varepsilon_{v,i,j,\pm}$ *given by the following formulas:*

(a) $\delta_{v,i,j,\pm} = \min_{i \leq k \leq j} \delta_{k,\pm}$,

(b) $G_{v,i,j,\pm} = \prod_{i \leq k \leq j}(F_{k,\pm}(L+1) + G_{k,\pm})$,
where
$F_{k,\pm} = \max_{l_{r,\pm} \leq r \leq L + l_{r,\pm}} |a_{k,\pm}[r]|$, *for* $0 \leq r \leq N$.

(c) $\varepsilon_{v,i,j,\pm} = \min_{i \leq k \leq j} \varepsilon_{k,\pm}$.

Proof. The asymptotic expansions with explicit upper bounds for remainders for probabilities $p_{i,\pm}(\varepsilon), 0 \leq i \leq N$ appearing in condition $\mathbf{M}'_L$ let us apply proposition

(ii) (the multiple multiplication rule with simplified upper bound for remainders) of Lemma 2.7 to the pivotal $(l_{i,j,\pm}, L + l_{i,j,\pm})$-expansion for function $V_{i,j,\pm}(\varepsilon)$ given in Lemma 5.2 and Remark 5.1. This directly yields relations given in propositions **(a)**–**(c)** of the Lemma 5.8. □

Lemma 5.9. *Let conditions* **J**–**L** *and* $\mathbf{M}'_L$ *hold. Then, for every* $0 \leq i, j \leq N$, *the* $(m_{i,j}, L + m_{i,j})$*-expansion for expectation* $U_{i,j}(\varepsilon)$ *given in Lemma 5.3 is a* $(m_{i,j}, L + m_{i,j}, \delta_{u,i,j}, G_{u,i,j}, \varepsilon_{u,i,j})$*-expansion, with parameters* $\delta_{u,i,j}, G_{u,i,j}$, *and* $\varepsilon_{u,i,j}$ *given by the following formulas:*

$$(\mathbf{a})\ \delta_{u,i,j} = \begin{cases} \delta_{v,j+1,i,-} \wedge \delta_{v,j,i-1,+} & \text{for } 0 \leq j < i \leq N, \\ 1 & \text{for } 0 \leq j = i \leq N, \\ \delta_{v,i,j-1,+} \wedge \delta_{v,+1,j,-} & \text{for } 0 \leq i < j \leq N, \end{cases}$$

$$(\mathbf{b})\ G_{u,i,j} = \begin{cases} \frac{2(G_{v,j+1,i,-} + F_{v,j+1,i,-} F_{u,i,j} (L+1)^2 + G_{v,j,i-1,+} F_{u,i,j} (L+1))}{v_{j,i-1,+}[l_{j,i-1,+}]} & \text{for } 0 \leq j < i \leq N, \\ 0 & \text{for } 0 \leq j = i \leq N, \\ \frac{2(G_{v,i,j-1,+} + F_{v,i,j-1,+} F_{u,i,j} (L+1)^2 + G_{v,i+1,j,-} F_{u,i,j} (L+1))}{v_{i+1,j,-}[l_{i+1,j,-}]} & \text{for } 0 \leq i < j \leq N, \end{cases}$$

where

$F_{v,i,j,\pm} = \max_{l_{i,j,\pm} \leq k \leq L + l_{i,j,\pm}} |v_{i,j,\pm}[k]|$, *for* $0 \leq i \leq j \leq N$,

and

$F_{u,i,j} = \max_{m_{i,j} \leq k \leq L + m_{i,j}} |u_{i,j}[k]|$, *for* $0 \leq i, j \leq N$.

$$(\mathbf{c})\ \varepsilon_{u,i,j} = \begin{cases} \varepsilon_{v,j+1,i,-} \wedge \varepsilon_{v,j,i-1,+} \wedge \tilde{\varepsilon}_{v,j,i-1,+} & \text{for } 0 \leq j < i \leq N, \\ \varepsilon_0 & \text{for } 0 \leq j = i \leq N, \\ \varepsilon_{v,i,j-1,+} \wedge \varepsilon_{v,i+1,j,-} \wedge \tilde{\varepsilon}_{v,i+1,j,-} & \text{for } 0 \leq i < j \leq N, \end{cases}$$

where

$\tilde{\varepsilon}_{v,i,j,\pm} = \left(\frac{v_{i,j,\pm}[l_{i,j,\pm}]}{2(F_{v,i,j,\pm} L + G_{v,i,j,\pm})}\right)^{\frac{1}{\delta_{v,i,j,\pm}}}$, *for* $0 \leq i \leq j \leq N$.

Proof. The asymptotic expansions with explicit upper bounds for remainders for products $V_{i,j,\pm}(\varepsilon), 0 \leq i \leq j \leq N$ given in Lemma 5.8 let us apply proposition **(v)** (the division rule with simplified upper bound for remainders) of Lemma 2.6 to the pivotal $(m_{i,j}, L + m_{i,j})$-expansions for functions $U_{i,j}(\varepsilon), 0 \leq i, j \leq N$ given in Lemma 5.3. This directly yields relations given in propositions **(a)**–**(c)** of Lemma 5.9. □

Lemma 5.10. *Let conditions* **J**–**L** *and* $\mathbf{N}'_L$ *hold. Then, for every* $0 \leq i \leq N$, *the pivotal* $(0, d)$*-expansion for expectation* $e_i(\varepsilon)$ *given in Lemma 5.4 is a pivotal* $(0, L, \dot{\delta}_i, \dot{G}_i, \dot{\varepsilon}_i)$*-expansion, with parameters* $\dot{\delta}_i, \dot{G}_i$ *and* $\dot{\varepsilon}_i$ *given by the following formulas,*

(a) $\dot{\delta}_i = \dot{\delta}_{i,-} \wedge \dot{\delta}_{i,+}$,

(b) $\dot{G}_i = \dot{F}_{i,-} + \dot{F}_{i,+} + \dot{G}_{i,-} + \dot{G}_{i,+}$,

where $\dot{F}_{i,\pm} = \max_{l_{i,\pm} \leq k \leq L+l_{i,\pm}} |b_{i,\pm}[k]|$, *for* $0 \leq i \leq N$.

(c) $\dot{\varepsilon}_i = \dot{\varepsilon}_{i,-} \wedge \dot{\varepsilon}_{i,+}$.

Proof. The asymptotic expansions with explicit upper bounds for remainders for expectations $e_{i,\pm}(\varepsilon), 0 \leq i \leq N$ appearing in condition $\mathbf{N}'_L$ let us apply proposition **(ii)** (the summation rule with simplified upper bound for remainders) of Lemma 2.6 to the pivotal $(0,L)$-expansions for expectations $e_i(\varepsilon), 0 \leq i \leq N$ given in Lemma 5.4. This directly yields relations given in propositions **(a)**–**(c)** of Lemma 5.10. □

Lemma 5.11. *Let conditions* **J–L**, $\mathbf{M}'_L$, $\mathbf{N}'_L$, *and* **P** *hold. Then, for every* $0 \leq i,j \leq N$, *the pivotal* $(m_{i,j}, L+m_{i,j})$*-expansion for expectation* $W_{i,j}(\varepsilon)$ *given in Lemma 5.5 is a pivotal* $(m_{i,j}, L+m_{i,j}, \delta_i, G_i, \varepsilon_i)$*-expansion, with parameters* δ_i, G_i, *and* ε_i *given by the following formulas:*

(a) $\delta_{w,i,j} = \dot{\delta}_j \wedge \delta_{u,i,j}$,

(b) $G_{w,i,j} = (\dot{F}_j(L+1) + \dot{G}_j)(F_{u,i,j}(L+1) + G_{u,i,j})$,

where $\dot{F}_i = \max_{0 \leq k \leq L} |b_i[k]|$,

(c) $\varepsilon_{w,i,j} = \dot{\varepsilon}_j \wedge \varepsilon_{u,i,j}$.

Proof. The asymptotic expansions with explicit upper bounds for remainders for expectations $e_i(\varepsilon), 0 \leq i \leq N$ and $U_{\varepsilon,i,j}, , 0 \leq i,j \leq N$ given, respectively, in Lemmas 5.9 and 5.10 let us apply proposition **(iii)** (the multiplication rule with simplified upper bound for remainders) of Lemma 2.6 to the pivotal $(m_{i,j}, L+m_{i,j})$-expansions for expectations $W_{i,j}(\varepsilon), 0 \leq i,j \leq N$ given in Lemma 5.5. This directly yields relations given in propositions **(a)**–**(c)** of Lemma 5.11. □

Lemma 5.12. *Let conditions* **J–L**, $\mathbf{M}'_L$, $\mathbf{N}'_L$, *and* **P** *hold. Then, for every* $0 \leq i,j \leq N$, *the pivotal* $(M_{ii}, L+M_{ii})$*-expansion for expectation of return time* $E_{ii}(\varepsilon)$ *given in Lemma 5.6 is a pivotal* $(M_{ii}, L+M_{ii}, \dot{\delta}_{ii}, \dot{G}_{ii}, \dot{\varepsilon}_{ii})$*-expansion, with parameters* $\dot{\delta}_{ii}, \dot{G}_{ii}$, *and* $\dot{\varepsilon}_{ii}$ *given by the following formulas:*

(a) $\dot{\delta}_{ii} = \min_{0 \leq j \leq N} \dot{\delta}_{w,i,j}$,

(b) $\dot{G}_{ii} = \sum_{0 \leq j \leq N} (\dot{F}_{w,i,j} + \dot{G}_{w,i,j})$,

where $\dot{F}_{w,i,j} = \max_{m_{i,j} \leq k \leq L+m_{i,j}} |w_{i,j}[k]|$,

(c) $\dot{\varepsilon}_{ii} = \min_{0 \leq j \leq N} \dot{\varepsilon}_{w,i,j}$.

Proof. The asymptotic expansions with explicit upper bounds for remainders for expectations $W_{i,j}(\varepsilon), 0 \leq i,j \leq N$ given in Lemma 5.11 let us apply proposition **(i)** (the multiple summation rule with simplified upper bound for remainders) of Lemma 2.7 to the pivotal $(M_{ii}, L+M_{ii})$-expansions for expectations $E_{ii}(\varepsilon), 0 \leq i \leq N$ given in Lemma 5.6. This directly yields relations given in propositions **(a)**–**(c)** of Lemma 5.11. □

5.3.2 Asymptotic Expansions with Explicit Upper Bounds for Remainders for Stationary and Conditional Quasi-Stationary Distributions

Theorem 5.3. *Let conditions* **J–L**, $\mathbf{M}'_L$, $\mathbf{N}'_L$, *and* **P** *hold. Then, for every* $0 \leq i \leq N$, *the pivotal* $(n_i, L+n_i)$*-expansion for stationary probability* $\pi_i(\varepsilon)$ *given in Theorem 5.1 is a pivotal* $(n_i, L+n_i, \delta_i, G_i, \varepsilon_i)$*-expansion, with parameters* δ_i, G_i, *and* ε_i *given by the following formulas:*

(a) $\delta_i = \dot{\delta}_i \wedge \dot{\delta}_{ii}$,

(b) $G_i = \frac{2(\dot{G}_i + \dot{F}_{ii}F_i(L+1)^2 + \dot{G}_{ii}F_i(L+1))}{B_{ii}[M_{ii}]}$,

where

$F_i = \max_{n_i \leq k \leq L+n_i} |c_i[k]|$, *for* $0 \leq i \leq N$.

(c) $\varepsilon_i = \dot{\varepsilon}_i \wedge \dot{\varepsilon}_{ii} \wedge \tilde{\varepsilon}_{ii}$,

where

$\tilde{\varepsilon}_{ii} = \left(\frac{B_{ii}[M_{ii}]}{2(\dot{F}_{ii}L + \dot{G}_{ii})}\right)^{\frac{1}{\delta_{ii}}}$,

and

$\dot{F}_{ii} = \max_{M_{ii} \leq k \leq L+M_{ii}} |B_{ii}[k]|$.

Proof. The asymptotic expansions with explicit upper bounds for remainders for expectations $e_i(\varepsilon), E_{ii}(\varepsilon), 0 \leq i \leq N$ given in Lemmas 5.10 and 5.12 let us apply proposition **(v)** (the division rule with simplified upper bound for remainders) of Lemma 2.6 to the pivotal $(n_i, L+n_i)$-expansions for stationary probabilities $\pi_i(\varepsilon), 0 \leq i \leq N$ given in Theorem 5.1. This directly yields relations given in propositions **(a)**–**(c)** of Theorem 5.3. □

Lemma 5.13. *Let conditions* **J–L**, $\mathbf{M}'_L$, $\mathbf{N}'_L$, *and* **P** *hold. Then, for* $r = 2, 3$, *the pivotal* $(1, L+1)$*-expansion for function* $\Pi_{\mathbb{U}_r}(\varepsilon)$ *given in Lemma 5.7 is a pivotal* $(1, L+1, \delta_{\mathbb{U}_r}, G_{\mathbb{U}_r}, \varepsilon_{\mathbb{U}_r})$*-expansion, with parameters* $\delta_{\mathbb{U}_r}, G_{\mathbb{U}_r}$, *and* $\varepsilon_{\mathbb{U}_r}$ *given by the following formulas:*

(a) $\delta_{\mathbb{U}_r} = \min_{i \in \mathbb{U}_r} \delta_i$,

(b) $G_{\mathbb{U}_r} = \sum_{i \in \mathbb{U}_r} (F_i + G_i)$,

(c) $\varepsilon_{\mathbb{U}_r} = \min_{i \in \mathbb{U}_r} \varepsilon_i$.

Proof. The asymptotic expansions with explicit upper bounds for remainders for stationary probabilities $\pi_i(\varepsilon), 0 \leq i \leq N$ given in Theorem 5.3 let us apply proposition **(i)** (the multiple summation rule with simplified upper bound for remainders) of Lemma 2.7 to pivotal $(1, L+1)$-expansions for functions $\Pi_{\mathbb{U}_r}(\varepsilon), r = 2, 3$ given in Lemma 5.7. This directly yields relations given in propositions **(a)**–**(c)** of Lemma 5.13. □

Theorem 5.4. *Let conditions* **J–L**, $\mathbf{M}'_L$, $\mathbf{N}'_L$, *and* **P** *hold. Then, for every* $i \in \mathbb{U}_r$, $r = 2, 3$, *the pivotal* $(0, d)$*-expansion for the conditional quasi-stationary probability*

$\pi_{\mathbb{U}_r,i}(\varepsilon)$ given in Theorem 5.2 is a pivotal $(0, L, \delta_{\mathbb{U}_r,i}, G_{\mathbb{U}_r,i}, \varepsilon_{\mathbb{U}_r,i})$-expansion, with parameters $\delta_{\mathbb{U}_r,i}, G_{\mathbb{U}_r,i}$, and $\varepsilon_{\mathbb{U}_r,i}$ given by the following formulas:

(a) $\delta_{\mathbb{U}_r,i} = \delta_i \wedge \delta_{\mathbb{U}_r}$,

(b) $G_{\mathbb{U}_r,i} = \frac{2(G_i + F_{\mathbb{U}_r} F_{\mathbb{U}_r,i}(L+1)^2 + G_{\mathbb{U}_r} F_{\mathbb{U}_r,i}(L+1))}{C_{\mathbb{U}_r}[1]}$,

where

$$F_{\mathbb{U}_r,i} = \max_{0 \leq k \leq L} |c_{\mathbb{U}_r,i}[k]|, \text{ for } r = 2, 3.$$

(c) $\varepsilon_{\mathbb{U}_r,i} = \varepsilon_i \wedge \varepsilon_{\mathbb{U}_r} \wedge \tilde{\varepsilon}_{\mathbb{U}_r}$, *where*

$$\tilde{\varepsilon}_{\mathbb{U}_r} = \Big(\frac{C_{\mathbb{U}_r}[1]}{2(F_{\mathbb{U}_r} L + G_{\mathbb{U}_r})}\Big)^{\frac{1}{\delta_{\mathbb{U}_r}}},$$

and

$$F_{\mathbb{U}_r} = \max_{1 \leq k \leq L+1} |C_{\mathbb{U}_r}[k]|.$$

Proof. The asymptotic expansions with explicit upper bounds for remainders for stationary probabilities $\pi_i(\varepsilon), i \in \mathbb{U}_r$ and functions $\Pi_{\mathbb{U}_r}(\varepsilon)$, for $r = 2, 3$, given, respectively, in Theorem 5.3 and Lemma 5.13, let us apply proposition **(v)** (the division rule with simplified upper bound for remainders) of Lemma 2.6 to the pivotal $(1, L+1)$-expansions for stationary probabilities $\pi_i(\varepsilon), i \in \mathbb{U}_r$ and functions $\Pi_{\mathbb{U}_r}(\varepsilon)$, given, respectively, in Theorem 5.1 and Lemma 5.7. This directly yields relations given in propositions **(a)**–**(c)** of Theorem 5.4. □

5.3.3 Asymptotic Expansions with Explicit Upper Bounds for Remainders and the Algorithm of Sequential Phase Space Reduction

The operational rules for Laurent asymptotic expansions formulated in Lemmas 2.1–2.8 can be applied to the recurrent formulas (5.14)–(5.16) used for computing transition probabilities and expectations of transition times for reduced perturbed birth-death semi-Markov processes.

As was pointed out in Subsection 5.3.2, the corresponding alternative recurrent algorithm would give the same Laurent asymptotic expansions for expectations $E_{ii}(\varepsilon), i \in \mathbb{X}$, as the algorithm presented in Lemmas 5.2–5.6. The only difference would be in forms of formulas for parameters, coefficients, and remainders in the corresponding expansions. However, both algorithms give the same value of parameter $\dot{\delta}_{ii}$ but can give different values for parameters $\dot{G}_{ii}$ and $\dot{\varepsilon}_{ii}$.

As far as asymptotic expansions with explicit upper bounds for remainders for stationary and quasi-stationary probabilities are concerned, they can be obtained using the algorithms described in Theorems 5.1–5.4, i.e., by application of the proposition **(ii)** (the division rule) given in Lemmas 2.2 and 2.6 to the corresponding quotients defining the stationary and quasi-stationary probabilities.

Chapter 6
Examples and Survey of Applied Perturbed Stochastic Models

In Chapter 6, we present a background for construction of numerical examples for asymptotic results presented in Chapters 2–5, some numerical examples as well as a summary survey of applied perturbed stochastic systems.

6.1 A Background for Numerical Examples

In this section we discuss some general questions related to construction of examples illustrating the asymptotic results for nonlinearly perturbed semi-Markov processes presented in Chapters 2–5.

6.1.1 Perturbation Conditions for Transition Probabilities of Embedded Markov Chains

Let $\mathbb{Y}_i \neq \emptyset, i \in \mathbb{X}$ be some subsets of space $\mathbb{X}$ such that condition **A (c)** holds for these sets, i.e., for every pair of states $i, j \in \mathbb{X}$, there exist an integer $n_{ij} \geq 1$ and a chain of states $i = l_{ij,0}, l_{ij,1}, \ldots, l_{ij,n_{ij}} = j$ such that $l_{ij,1} \in \mathbb{Y}_{l_{ij,0}}, \ldots, l_{ij,n_{ij}} \in \mathbb{Y}_{l_{ij,n_{ij}-1}}$.

Let us also choose some $\varepsilon_0 \in (0,1]$. We define $p_{ij}(\varepsilon) = 0, \varepsilon \in (0, \varepsilon_0], j \in \overline{\mathbb{Y}}_i, i \in \mathbb{X}$, i.e., assume that condition **A (b)** holds.

Let $p_{ij}(\varepsilon), \varepsilon \in (0, \varepsilon_0], j \in \mathbb{Y}_i, i \in \mathbb{X}$ be some real-valued functions, which satisfy condition **D**, i.e., can be represented in the form of Taylor asymptotic expansions, $p_{ij}(\varepsilon) = \sum_{l=l_{ij}^-}^{l_{ij}^+} a_{ij}[l]\varepsilon^l + o_{ij}(\varepsilon^{l_{ij}^+}), \varepsilon \in (0, \varepsilon_0]$, for $j \in \mathbb{Y}_i, i \in \mathbb{X}$, where, for $j \in \mathbb{Y}_i, i \in \mathbb{X}$, **(a)** $a_{ij}[l_{ij}^-] > 0$ and $0 \leq l_{ij}^- \leq l_{ij}^+ < \infty$; **(b)** $o_{ij}(\varepsilon^{l_{ij}^+})/\varepsilon^{l_{ij}^+} \to 0$ as $\varepsilon \to 0$.

D. Silvestrov, S. Silvestrov, *Nonlinearly Perturbed Semi-Markov Processes*,
SpringerBriefs in Probability and Mathematical Statistics,
DOI 10.1007/978-3-319-60988-1_6

Condition **D** does not guarantee that matrix $\|p_{ij}(\varepsilon)\|$ is stochastic, for every $\varepsilon \in (0,\varepsilon_0]$. This can be achieved by imposing some additional conditions on coefficients and remainders in the above asymptotic expansions.

Let us choose arbitrary numbers $0 < \alpha_{ij} < \frac{1}{2}$, for $j \in \mathbb{Y}_i, i \in \mathbb{X}$.

Condition **D (b)** guarantees that, for every $j \in \mathbb{Y}_i$, $i \in \mathbb{X}$, there exists $\varepsilon_{\alpha_{ij},ij} \in (0,\varepsilon_0]$ such that, $|o_{ij}(\varepsilon^{l^+_{ij}})/\varepsilon^{l^+_{ij}}| \leq \alpha_{ij}|a_{ij}[l^-_{ij}]|$, for $\varepsilon \in (0,\varepsilon_{\alpha_{ij},ij}]$.

It is useful to note that in the case, where condition **D**$'$ holds, an explicit value for parameters $\varepsilon_{\alpha_{ij},ij}$ can be derived from the inequalities, $|o_{ij}(\varepsilon^{l^+_{ij}})| \leq G_{ij}\varepsilon^{l^+_{ij}+\delta_{ij}}, \varepsilon \in (0,\varepsilon_{ij}], j \in \mathbb{Y}_i$, $i \in \mathbb{X}$, appearing in this condition. In this case, one can take $\varepsilon_{\alpha,ij} = \varepsilon_{ij} \wedge (\frac{\alpha_{ij}|a_{ij}[l^-_{ij}]|}{G_{ij}})^{\frac{1}{\delta_{ij}}}, j \in \mathbb{Y}_i$, $i \in \mathbb{X}$.

Let us also define $A_{\varepsilon_0,ij} = \sum_{l^-_{ij} < l \leq l^+_{ij}} |a_{ij}[l]|\varepsilon_0^{l-l^-_{ij}-1}, j \in \mathbb{Y}_i, i \in \mathbb{X}$ and $\varepsilon'_{\alpha_{ij},ij} = \varepsilon_{\alpha_{ij},ij}$, if $A_{\varepsilon_0,ij} = 0$, or $\varepsilon'_{\alpha_{ij},ij} = \varepsilon_{\alpha_{ij},ij} \wedge \frac{\alpha_{ij}|a_{ij}[l^-_{ij}]|}{A_{\varepsilon_0,ij}}$, if $A_{\varepsilon_0,ij} > 0$. Inequalities, $p_{ij}(\varepsilon) \geq \varepsilon^{l^-_{ij}}(a_{ij}[l^-_{ij}] - \varepsilon A_{\varepsilon_0,ij} - \varepsilon_0^{l^+_{ij}-l^-_{ij}}|o_{ij}(\varepsilon^{l^+_{ij}})/\varepsilon^{l^+_{ij}}|) \geq \varepsilon^{l^-_{ij}} a_{ij}[l^-_{ij}](1-2\alpha_{ij}) > 0$, hold, for every $\varepsilon \in (0,\varepsilon'_{\alpha_{ij},ij}]$ and $j \in \mathbb{Y}_i$, $i \in \mathbb{X}$,

Let us now define $\varepsilon'_0 = \min_{j\in\mathbb{Y}_i, i\in\mathbb{X}} \varepsilon'_{\alpha_{ij},ij}$.

Obviously, $p_{ij}(\varepsilon) > 0$, for $j \in \mathbb{Y}_i$, $i \in \mathbb{X}$ and $\varepsilon \in (0,\varepsilon'_0]$. Thus, conditions **A (a)**–**(b)** hold, if parameter ε_0 is replaced by the new value ε'_0.

The question about holding of the stochasticity relation $\sum_{j\in\mathbb{Y}_i} p_{ij}(\varepsilon) = 1, \varepsilon \in (0,\varepsilon_0], i \in \mathbb{X}$ is more complex. According to Lemma 3.1, under conditions **A (a)**–**(b)** and **D**, the above stochasticity relation is equivalent to condition **F** formulated in Section 3.1.

First, condition **F** requires holding of the following relation:

$$\sum_{j\in\mathbb{Y}_i} a_{ij}[l] = \mathrm{I}(l=0),\ 0 \leq l \leq l^+_{i,\mathbb{Y}_i},\ i \in \mathbb{X}, \tag{6.1}$$

where (a) $l^{\pm}_{i,\mathbb{Y}_i} = \min_{j\in\mathbb{Y}_i} l^{\pm}_{ij}, i \in \mathbb{X}$ and (b) $a_{ij}[l] = 0$, for $0 \leq l < l^-_{ij}, j \in \mathbb{Y}_i, i \in \mathbb{X}$.

Note that parameters $l^-_{i,\mathbb{Y}_i} = 0, i \in \mathbb{X}$.

It is not difficult to choose coefficients $a_{ij}[l], l = l^-_{ij} \leq l \leq l^+_{ij}, j \in \mathbb{Y}_i$, $i \in \mathbb{X}$ in such way that relation (6.1) would hold. Any such coefficients, with the first coefficients $a_{ij}[l^-_{ij}] > 0, j \in \mathbb{Y}_i$, $i \in \mathbb{X}$, can serve as coefficients in the asymptotic expansions appearing in condition **D**.

Second, condition **F** requires holding of identity, for every $i \in \mathbb{X}$,

$$\sum_{j\in\mathbb{Y}_i}\Big(\sum_{l^+_{i,\mathbb{Y}_i} < l \leq l^+_{ij}} a_{ij}[l]\varepsilon^l + o_{ij}(\varepsilon^{l^+_{ij}})\Big) \equiv 0. \tag{6.2}$$

Remainders $o_{ij}(\varepsilon^{l^+_{ij}}), j \in \mathbb{Y}_i$, $i \in \mathbb{X}$ satisfying the above identities can be chosen in different ways.

The simplest one is to choose $o_{ij}(\varepsilon^{l^+_{ij}}) \equiv 0$, $j \in \mathbb{Y}_i$, $i \in \mathbb{X}$. In this case, the above identities would reduce to equalities, $\sum_{j\in\mathbb{Y}_i} a_{ij}[l] = 0, l^+_{i,\mathbb{Y}_i} < l \leq l^*_{i,\mathbb{Y}_i}, i \in \mathbb{X}$, where

(a) $l^*_{i,\mathbb{Y}_i} = \max_{j\in\mathbb{Y}_i} l^+_{ij}, i \in \mathbb{X}$ and (b) $a_{ij}[l] = 0$, for $l^+_{ij} < l \leq l^*_{i,\mathbb{Y}_i}, j \in \mathbb{Y}_i, i \in \mathbb{X}$. These equalities supplement equalities given in relation (6.1). Such choice of remainders corresponds to models with polynomial perturbations.

We, however, would like to impose on remainders conditions mainly required of them by conditions **D** or **D**$'$.

There always exist $j_i \in \mathbb{Y}_i, i \in \mathbb{X}$ such that $l^+_{ij_i} = l^+_{i,\mathbb{Y}_i}, i \in \mathbb{X}$.

Identity (6.2) can be rewritten in the following form, for every $i \in \mathbb{X}$:

$$o_{ij_i}(\varepsilon^{l^+_{ij_i}}) \equiv - \sum_{j\in\mathbb{Y}_i, j\neq j_i} \Big(\sum_{l^+_{i,\mathbb{Y}_i} < l \leq l^+_{ij}} a_{ij}[l]\varepsilon^l + o_{ij}(\varepsilon^{l^+_{ij}})\Big). \tag{6.3}$$

Relation (6.3) can be used as the formula defining remainders $o_{ij_i}(\varepsilon^{l^+_{ij_i}}), i \in \mathbb{X}$, via remainders $o_{ij}(\varepsilon^{l^+_{ij}}), j \in \mathbb{Y}_i, j \neq j_i, i \in \mathbb{X}$ appearing in the corresponding asymptotic expansions in condition **D**.

Since $l^+_{ij_i} = l^+_{i,\mathbb{Y}_i}, i \in \mathbb{X}$, the relation, $o_{ij_i}(\varepsilon^{l^+_{ij_i}})/\varepsilon^{l^+_{ij_i}} \to 0$ as $\varepsilon \to 0$, holds, for every $i \in \mathbb{X}$, for remainder $o_{ij}(\varepsilon^{l^+_{ij}})$ defined by relation (6.3).

Thus, remainders $o_{ij_i}(\varepsilon^{l^+_{ij_i}}), i \in \mathbb{X}$ defined by relation (6.3) can also serve in the corresponding asymptotic expansions in condition **D**.

Moreover, let us assume that remainders $o_{ij}(\varepsilon^{l^+_{ij}}), j \in \mathbb{Y}_i, j \neq j_i, i \in \mathbb{X}$ satisfy the inequalities, $|o_{ij}(\varepsilon^{l^+_{ij}})| \leq G_{ij}\varepsilon^{l^+_{ij}+\delta_{ij}}, \varepsilon \in (0, \varepsilon_{ij}], j \in \mathbb{Y}_i, j \neq i_j, i \in \mathbb{X}$, appearing in condition **D**$'$.

Let us define $\varepsilon_{ij_i} = \min_{j\in\mathbb{Y}_i, j\neq j_i} \varepsilon_{ij}, i \in \mathbb{X}$ and $\delta_{ij_i} = \min_{j\in\mathbb{Y}_i, j\neq j_i} \delta_{ij}, i \in \mathbb{X}$.

In this case, inequalities, $|o_{ij_i}(\varepsilon^{l^+_{ij_i}})| \leq \sum_{j\in\mathbb{Y}_i, j\neq j_i}(\sum_{l^+_{i,\mathbb{Y}_i} < l \leq l^+_{ij}} |a_{ij}[l]|\varepsilon^l + |o_{ij}(\varepsilon^{l^+_{ij}})|)$ $\leq \big(\sum_{j\in\mathbb{Y}_i, j\neq j_i}(\sum_{l^+_{i,\mathbb{Y}_i} < l \leq l^+_{ij}} |a_{ij}[l]|\,\varepsilon_0^{l - l_{ij_i} - \delta_{ij_i}} + \varepsilon_0^{l^+_{ij} - l^+_{ij_i} + \delta_{ij} - \delta_{ij_i}} G_{ij})\big)\varepsilon^{l^+_{ij_i}+\delta_{ij_i}} = G_{ij_i}\varepsilon^{l^+_{ij_i}+\delta_{ij_i}}$, hold, for every $\varepsilon \in (0, \varepsilon_{ij_i}], i \in \mathbb{X}$,

Thus, the inequalities, $|o_{ij_i}(\varepsilon^{l^+_{ij_i}})| \leq G_{ij_i}\varepsilon^{l^+_{ij_i}+\delta_{ij_i}}, \varepsilon \in (0, \varepsilon_{ij_i}], i \in \mathbb{X}$, appearing in condition **D**$'$ hold for remainders $o_{ij_i}(\varepsilon^{l^+_{ij_i}}), i \in \mathbb{X}$, with parameters $\varepsilon_{ij_i}, \delta_{ij_i}$, and G_{ij_i} defined above.

As follows from the above remarks, identity (6.3) holds for remainders $o_{ij}(\varepsilon^{l^+_{ij}})$, $j \in \mathbb{Y}_i, , i \in \mathbb{X}$, for $\varepsilon \in (0, \varepsilon''_0], i \in \mathbb{X}$, where $\varepsilon''_0 = \min_{j\in\mathbb{Y}_i} \varepsilon_{ij} = \min_{j\in\mathbb{Y}_i, j\neq j_i} \varepsilon_{ij}$. Thus, condition **F** holds, if parameter ε_0 is replaced by the new value ε''_0.

By the above remarks, functions $p_{ij}(\varepsilon), i, j \in \mathbb{X}$ can, for every $\varepsilon \in (0, \varepsilon'_0 \wedge \varepsilon''_0]$, serve as transition probabilities of a Markov chain.

Note that remainders $o_{ij}(\varepsilon^{l^+_{ij}}), j \in \mathbb{Y}_i, i \in \mathbb{X}$ constructed above can be very irregular functions. Let us, for example, consider the case, where all asymptotic expansions in condition **D** have the same order, i.e., parameters $l^+_{ij} = l^+, j \in \mathbb{Y}_i, i \in \mathbb{X}$. In this case, identities (6.2) take the form, $\sum_{j\in\mathbb{Y}_i} o_{ij}(\varepsilon^{l^+}) = 0, \varepsilon \in (0, \varepsilon_0], i \in \mathbb{X}$. Condition **D** requires that $o_{ij}(\varepsilon^{l^+})/\varepsilon^{l^+} \to 0$ as $\varepsilon \to 0$, for $j \in \mathbb{Y}_i, i \in \mathbb{X}$. Remainders $o_{ij}(\varepsilon^{l^+}), j \in \mathbb{Y}_i, i \in \mathbb{X}$ can be continuous functions of ε taking zero value in at most finite numbers of points. However, let us multiply them, for example, by the Dirichlet function $D(\varepsilon)$. The new remainders $o'_{ij}(\varepsilon^{l^+}) = D(\varepsilon)o_{ij}(\varepsilon^{l^+}), j \in \mathbb{Y}_i, i \in \mathbb{X}$ also satisfy identities (6.2) and $o'_{ij}(\varepsilon^{l^+})/\varepsilon^{l^+} \to 0$ as $\varepsilon \to 0$, for $j \in \mathbb{Y}_i, i \in \mathbb{X}$. At the same time, they are very irregular functions. This example is, of course, an artificial one. But, it well illustrates the above statement about possible irregularity of remainders and, in sequel, transition probabilities, as functions of the perturbation parameter.

6.1.2 Perturbation Conditions for Expectations of Transition Times

Let us also make some remarks concerned the expected transition times.

First, let us define $e_{ij}(1,\varepsilon) = 0, \varepsilon \in (0,\varepsilon_0]\ j \in \overline{\mathbb{Y}}_i,\ i \in \mathbb{X}$ that is consistent with condition **A (b)**.

Let us also $e_{ij}(1,\varepsilon), \varepsilon \in (0,\varepsilon_0]\ j \in \mathbb{Y}_i,\ i \in \mathbb{X}$ be some real-valued functions which satisfy condition $\mathbf{E}_1$, i.e., can be represented in the form of Laurent asymptotic expansions, $e_{ij}(1,\varepsilon) = \sum_{l=m^-_{ij}[1]}^{m^+_{ij}[1]} b_{ij}[1,l]\varepsilon^l + \dot{o}_{1,ij}(\varepsilon^{m^+_{ij}[1]})$, $\varepsilon \in (0,\varepsilon_0]$, for $j \in \mathbb{Y}_i, i \in \mathbb{X}$, where **(a)** $b_{ij}[1,m^-_{ij}[1]] > 0$ and $-\infty < m^-_{ij}[1] \leq m^+_{ij}[1] < \infty$, for $j \in \mathbb{Y}_i, i \in \mathbb{X}$; **(b)** $\dot{o}_{1,ij}(\varepsilon^{m^+_{ij}[1]})/\varepsilon^{m^+_{ij}[1]} \to 0$ as $\varepsilon \to 0$, for $j \in \mathbb{Y}_i, i \in \mathbb{X}$.

It is useful noting again that periods above the letters denoting, parameters and remainders appear in condition $\mathbf{E}'_1$ in order to distingue these parameters and remainders of parameters and remainders in condition $\mathbf{D}'$.

Condition $\mathbf{E}_1$ **(b)** guarantees that, for every $j \in \mathbb{Y}_i$, $i \in \mathbb{X}$ there exists $\dot{\varepsilon}_{\alpha_{ij},ij} \in (0,\varepsilon_0]$ such that, $|\dot{o}_{1,ij}(\varepsilon^{m^+_{ij}[1]})/\varepsilon^{m^+_{ij}[1]}| \leq \alpha_{ij}|b_{ij}[1,m^-_{ij}[1]]|$, for $\varepsilon \in (0,\dot{\varepsilon}_{\alpha_{ij},ij}]$.

It is useful to note that in the case, where condition $\mathbf{E}'_1$ is assumed to hold, an explicit value for parameters $\dot{\varepsilon}_{\alpha_{ij},ij}$ can be derived from the inequalities, $|\dot{o}_{1,ij}(\varepsilon^{m^+_{ij}[1]})| \leq \dot{G}_{ij}[1]\varepsilon^{m^+_{ij}[1]+\dot{\delta}_{ij}[1]}, \varepsilon \in (0,\dot{\varepsilon}_{ij}[1]], j \in \mathbb{Y}_i,\ i \in \mathbb{X}$, appearing in this condition. In this case, one can take $\dot{\varepsilon}_{\alpha,ij} = \dot{\varepsilon}_{ij}[1] \wedge (\frac{\alpha_{ij}|b_{ij}[1,m^-_{ij}[1]]|}{\dot{G}_{ij}[1]})^{\frac{1}{\dot{\delta}_{ij}[1]}}, j \in \mathbb{Y}_i,\ i \in \mathbb{X}$.

Let us also define $B_{\varepsilon_0,ij} = \sum_{m^-_{ij}[1] < l \leq m^+_{ij}[1]} |b_{ij}[1,l]|\varepsilon_0^{l-m^-_{ij}[1]-1}, j \in \mathbb{Y}_i,\ i \in \mathbb{X}$ and $\dot{\varepsilon}''_{\alpha_{ij},ij} = \dot{\varepsilon}_{\alpha_{ij},ij}$ if $B_{\varepsilon_0,ij} = 0$ or $\dot{\varepsilon}''_{\alpha_{ij},ij} = \dot{\varepsilon}_{\alpha_{ij},ij} \wedge \frac{\alpha_{ij}|b_{ij}[1,l^-_{ij}]|}{B_{\varepsilon_0,ij}}$ if $B_{\varepsilon_0,ij} > 0$. Inequalities, $e_{ij}(1,\varepsilon) \geq \varepsilon^{m^-_{ij}[1]}(b_{ij}[1,m^-_{ij}[1]] - \varepsilon B_{\varepsilon_0,ij} - \varepsilon_0^{m^+_{ij}[1]-m^-_{ij}[1]}|o_{1,ij}(\varepsilon^{m^+_{ij}[1]})/\varepsilon^{m^+_{ij}[1]}|) \geq \varepsilon^{m^-_{ij}[1]}b_{ij}[1,m^-_{ij}[1]](1-2\alpha_{ij}) > 0$, hold, for every $\varepsilon \in (0,\dot{\varepsilon}''_{\alpha_{ij},ij}]$ and $j \in \mathbb{Y}_i, i \in \mathbb{X}$.

Let us now define $\varepsilon'''_0 = \min_{j \in \mathbb{Y}_i, i \in \mathbb{X}} \dot{\varepsilon}''_{\alpha_{ij},ij}$.

Obviously, $e_{ij}(1,\varepsilon) > 0, j \in \mathbb{Y}_i,\ i \in \mathbb{X}, \varepsilon \in (0,\varepsilon'''_0]$. This is consistent, with conditions **A (a)** and **B**.

Finally, let us define $\tilde{\varepsilon}_0 = \varepsilon'_0 \wedge \varepsilon''_0 \wedge \varepsilon'''_0$. Parameter, $\tilde{\varepsilon}_0$ can serve as a new value for parameter ε_0.

Functions $p_{ij}(\varepsilon), i,j \in \mathbb{X}$ and $e_{ij}(1,\varepsilon), i,j \in \mathbb{X}$ constructed above can serve, respectively, as transition probabilities of the embedded Markov chain $\eta_{\varepsilon,n}$ and expectations of transition times for some semi-Markov process $\eta_\varepsilon(t)$, for every $\varepsilon \in (0, \tilde{\varepsilon}_0]$.

It is also worse to note that, under the assumption of holding condition **A (a)–(b)**, the perturbation conditions **D** and $\mathbf{E}_1$ are independent.

To see this, let us take arbitrary positive functions $p_{ij}(\varepsilon), j \in \mathbb{Y}_i, i \in \mathbb{X}$ and $e_{ij}(1,\varepsilon), j \in \mathbb{Y}_i, i \in \mathbb{X}$ satisfying, respectively, conditions **D** and $\mathbf{E}_1$, and, also, the corresponding relations (6.1) and (6.2). Then, there exist semi-Markov transition probabilities $Q_{\varepsilon,ij}(t), t \geq 0, j \in \mathbb{Y}_i, i \in \mathbb{X}$ such that $Q_{\varepsilon,ij}(\infty) = p_{ij}(\varepsilon), j \in \mathbb{Y}_i, i \in \mathbb{X}$ and $\int_0^\infty t\, Q_{\varepsilon,ij}(dt) = e_{ij}(1,\varepsilon), j \in \mathbb{Y}_i, i \in \mathbb{X}$, for every $\varepsilon \in (0, \varepsilon_0]$. It is readily seen that, for example, semi-Markov transition probabilities $Q_{\varepsilon,ij}(t) = \mathrm{I}(t \geq \frac{e_{ij}(1,\varepsilon)}{p_{ij}(\varepsilon)}) p_{ij}(\varepsilon), t \geq 0, j \in \mathbb{Y}_i, i \in \mathbb{X}$ satisfy the above relations.

6.2 Numerical Examples

In this section, we give numerical examples illustrating algorithms for construction of asymptotic expansions for nonlinearly perturbed semi-Markov processes.

6.2.1 Asymptotic Expansions with Remainders Given in the Standard Form

We assume that the semi-Markov process $\eta_\varepsilon(t)$ is, for every $\varepsilon \in (0, \varepsilon_0]$, a semi-Markov process with the phase space $\mathbb{X} = \{1,2,3\}$.

The transition sets are $\mathbb{Y}_1 = \{1,2\}, \mathbb{Y}_2 = \{1,2,3\}, \mathbb{Y}_3 = \{1,2\}$.

The 3×3 matrix of transition probabilities $\|p_{ij}(\varepsilon)\|$, for the corresponding embedded Markov chain $\eta_{\varepsilon,n}$, has the following form:

$$\left\| \begin{array}{lll} 1 + o_{11}(\varepsilon) & \varepsilon^2 + \varepsilon^3 + o_{12}(\varepsilon^3) & 0 \\ \frac{1}{2}\varepsilon + \frac{1}{2}\varepsilon^2 + o_{21}(\varepsilon^2) & 1 - \varepsilon + o_{22}(\varepsilon) & \frac{1}{2}\varepsilon + \frac{1}{2}\varepsilon^2 + o_{23}(\varepsilon^2) \\ \frac{1}{2} + o_{31}(\varepsilon) & \frac{1}{2} + o_{32}(\varepsilon) & 0 \end{array} \right\|. \tag{6.4}$$

The 3×3 matrix of expectations of sojourn times $\|e_{ij}(1,\varepsilon)\|$, for the semi-Markov process $\eta_\varepsilon(t)$, has the following form:

$$\left\| \begin{array}{lll} \varepsilon + \varepsilon^2 + \dot{o}_{1,11}(\varepsilon^2) & \varepsilon^3 + \varepsilon^4 + \dot{o}_{1,12}(\varepsilon^4) & 0 \\ \varepsilon + \varepsilon^2 + \dot{o}_{1,21}(\varepsilon^2) & 1 - \varepsilon + \dot{o}_{1,22}(\varepsilon) & 2 + \varepsilon + \dot{o}_{1,23}(\varepsilon) \\ \varepsilon^{-1} + 1 + \dot{o}_{1,31}(1) & 2\varepsilon^{-1} + \dot{o}_{1,32}(1) & 0 \end{array} \right\|. \tag{6.5}$$

In the asymptotic expansions appearing in relations (6.4) and (6.5), the coefficients $a_{ij}[l_{ij}^-], b_{ij}[1, m_{ij}^-[1]] > 0, j \in \mathbb{Y}_i, i \in \mathbb{X}$, and coefficients $a_{ij}[l], l = l_{ij}^-, \ldots, l_{ij}^+$, $j \in \mathbb{Y}_i, i \in \mathbb{X}$ satisfy relation (6.1). We also assume that parameter $\varepsilon_0 = \tilde{\varepsilon}_0$ and remainders $o_{ij}(\varepsilon^{l_{ij}^+}), \dot{o}_{1,ij}(\varepsilon^{m_{ij}^+[1]})$, $j \in \mathbb{Y}_i$, $i \in \mathbb{X}$, in the asymptotic expansions representing elements of matrices given in relations (6.4) and (6.5), are chosen according to the procedures described in Section 6.1.

In this case, matrices, given in the above relations, can, for every $\varepsilon \in (0, \varepsilon_0]$, serve as, respectively, the matrix of transition probabilities for the corresponding embedded Markov chain (these matrices are stochastic) and the matrix of expectations of transition times (all elements of these matrices are non-negative), for the semi-Markov process $\eta_\varepsilon(t)$, and conditions **A**, **B**, $\mathbf{C}_1$, **D**, $\mathbf{E}_1$, and **F** hold.

The matrices of transition probabilities $\|p_{ij}(\varepsilon)\|$, for the embedded Markov chains $\eta_{\varepsilon,n}$, and $\|p_{ij}(0)\|$, for the corresponding limiting Markov chain $\eta_{0,n}$, have, respectively, the following structures:

$$\left\| \begin{matrix} \bullet & \bullet & 0 \\ \bullet & \bullet & \bullet \\ \bullet & \bullet & 0 \end{matrix} \right\| \quad \text{and} \quad \left\| \begin{matrix} 1 & 0 & 0 \\ 0 & 1 & 0 \\ \frac{1}{2} & \frac{1}{2} & 0 \end{matrix} \right\|, \tag{6.6}$$

where symbol $\bullet$ indicates positions of positive elements in matrices $\|p_{ij}(\varepsilon)\|, \varepsilon \in (0, \varepsilon_0]$.

The phase space $\mathbb{X}$ is one class of communicative states for the embedded Markov chain $\eta_{\varepsilon,n}$, for every $\varepsilon \in (0, \varepsilon_0]$, while it consists of two closed classes of communicative states $\mathbb{X}_1 = \{1\}, \mathbb{X}_2 = \{2\}$ and the class of transient states $\mathbb{X}_3 = \{3\}$, for the limiting Markov chain $\eta_{0,n}$.

By excluding the state 1 from the phase space $\mathbb{X}$ and using the algorithms described in Sections 3.2 and 3.3, we construct the reduced semi-Markov processes ${}_1\eta_\varepsilon(t)$, with the phase space ${}_1\mathbb{X} = \{2, 3\}$. Conditions **A**, **B**, $\mathbf{C}_1$, **D**, $\mathbf{E}_1$, and **F** (with the same parameter ε_0) hold for these reduced semi-Markov processes. The corresponding transition sets are ${}_1\mathbb{Y}_2 = \{2, 3\}$ and ${}_1\mathbb{Y}_3 = \{2\}$. By applying the algorithms described in Theorems 3.2–3.4, we can compute the 2×2 matrices $\|{}_1p_{ij}(\varepsilon)\|$ and $\|{}_1e_{ij}(1, \varepsilon)\|$. These matrices take the following forms:

$$\left\| \begin{matrix} {}_1p_{22}(\varepsilon) & {}_1p_{23}(\varepsilon) \\ {}_1p_{32}(\varepsilon) & {}_1p_{33}(\varepsilon) \end{matrix} \right\| = \left\| \begin{matrix} 1 - \frac{1}{2}\varepsilon + {}_1o_{22}(\varepsilon) & \frac{1}{2}\varepsilon + \frac{1}{2}\varepsilon^2 + {}_1o_{23}(\varepsilon^2) \\ 1 & 0 \end{matrix} \right\|, \tag{6.7}$$

and

$$\left\| \begin{matrix} {}_1e_{22}(1, \varepsilon) & {}_1e_{23}(1, \varepsilon) \\ {}_1e_{32}(1, \varepsilon) & {}_1e_{33}(1, \varepsilon) \end{matrix} \right\| = \left\| \begin{matrix} \frac{3}{2} + \frac{1}{2}\varepsilon + {}_1\dot{o}_{1,22}(\varepsilon) & 2 + \varepsilon + {}_1\dot{o}_{1,23}(\varepsilon) \\ \frac{7}{2}\varepsilon^{-1} + 1 + {}_1\dot{o}_{1,32}(1) & 0 \end{matrix} \right\|. \tag{6.8}$$

By excluding the state 2 from the reduced phase space ${}_1\mathbb{X} = \{2, 3\}$, we construct the "final" reduced semi-Markov processes ${}_{\langle 1,2 \rangle}\eta_\varepsilon(t)$, with the one-state phase space ${}_{\langle 1,2 \rangle}\mathbb{X} = \{3\}$. Conditions **A**, **B**, $\mathbf{C}_1$, **D**, $\mathbf{E}_1$, and **F** also hold for these semi-Markov processes. The corresponding transition set ${}_{\langle 1,2 \rangle}\mathbb{Y} = \{3\}$. The transition probability ${}_{\langle 1,2 \rangle}p_{33}(\varepsilon) \equiv 1$.

By applying the algorithms described in Theorems 3.3 and 3.4, we can compute the Laurent asymptotic expansion for the expected return time, ${}_{\langle 1,2 \rangle}e_{33}(1, \varepsilon) = E_{33}(1, \varepsilon) = \frac{21}{2}\varepsilon^{-1} - 3 + \ddot{o}_{1,33}(1)$.

The Laurent asymptotic expansion for the expected sojourn time $e_3(\varepsilon) = e_{31}(1,\varepsilon) + e_{32}(1,\varepsilon) + e_{33}(1,\varepsilon)$, obtained with the use of proposition **(ii)** (the multiple summation rule), given in Lemma 2.2, has the form, $e_3(\varepsilon) = 3\varepsilon^{-1} + 1 + \dot{o}_3(1)$.

Finally, the algorithm described in Theorem 4.2 gives the asymptotic expansion, for the stationary probability, $\pi_3(\varepsilon) = \frac{e_3(1,\varepsilon)}{E_{33}(1,\varepsilon)} = \frac{3\varepsilon^{-1}+1+\dot{o}_3(1)}{\frac{21}{2}\varepsilon^{-1}-3+\ddot{o}_{1,33}(1)} = \frac{2}{7} + \frac{26}{147}\varepsilon + o_3(\varepsilon)$.

Also, by excluding the state 3 from the reduced phase space ${}_1\mathbb{X} = \{2,3\}$ and applying the algorithms described in Theorems 3.2–3.4, we can compute the Laurent asymptotic expansion for the expected return time, ${}_{\langle 1,3\rangle}e_{22}(1,\varepsilon) = E_{22}(1,\varepsilon) = \frac{21}{4} + \frac{15}{4}\varepsilon + \ddot{o}_{1,22}(\varepsilon)$.

In this case, the asymptotic expansion for the expected sojourn time $e_2(\varepsilon) = 3 + \varepsilon + 2\varepsilon^2 + \dot{o}_2(\varepsilon^2)$, and the algorithm described in Theorem 4.2, gives the asymptotic expansion for the stationary probability, $\pi_2(\varepsilon) = \frac{e_2(\varepsilon)}{E_{22}(1,\varepsilon)} = \frac{3+\varepsilon+2\varepsilon^2+\dot{o}_2(\varepsilon^2)}{\frac{21}{4}+\frac{15}{4}\varepsilon+\ddot{o}_{1,22}(\varepsilon)} = \frac{4}{7} - \frac{32}{147}\varepsilon + o_2(\varepsilon)$.

As far as the stationary probability $\pi_1(\varepsilon)$ is concerned, the corresponding asymptotic expansion can be found using the identity, $\pi_1(\varepsilon) = 1 - \pi_2(\varepsilon) - \pi_3(\varepsilon), \varepsilon \in (0,\varepsilon_0]$, and the operational rules for asymptotic expansions given in Lemma 2.3. This yields the asymptotic expansion, $\pi_1(\varepsilon) = \frac{1}{7} + \frac{6}{147}\varepsilon + o_1(\varepsilon)$.

Alternatively, the exclusion of states from the phase space $\mathbb{X}$ in the opposite order, first state 3 and then state 2 or 1, and the use of the algorithms described in Theorems 3.2–3.4 yields the asymptotic expansions for the expected return times, ${}_{\langle 3,2\rangle}e_{11}(1,\varepsilon) = E_{11}(1,\varepsilon) = 7\varepsilon + \frac{15}{3}\varepsilon^2 + \ddot{o}_{1,11}(\varepsilon^2)$, and, ${}_{\langle 3,1\rangle}e_{22}(1,\varepsilon) = E_{22}(1,\varepsilon) = \frac{21}{4} + \frac{15}{4}\varepsilon + \ddot{o}_{1,22}(\varepsilon)$.

Then, the algorithm described in Theorem 4.2 yields the same asymptotic expansions for stationary probabilities, $\pi_1(\varepsilon) = \frac{e_1(\varepsilon)}{E_{11}(1,\varepsilon)} = \frac{\varepsilon+\varepsilon^2+\dot{o}_1(\varepsilon^2)}{7\varepsilon+\frac{15}{3}\varepsilon^2+\ddot{o}_{1,11}(\varepsilon^2)} = \frac{1}{7} + \frac{6}{147}\varepsilon + o_1(\varepsilon)$, and, $\pi_2(\varepsilon) = \frac{e_2(\varepsilon)}{E_{22}(1,\varepsilon)} = \frac{3+\varepsilon+2\varepsilon^2+\dot{o}_2(\varepsilon^2)}{\frac{21}{4}+\frac{15}{4}\varepsilon+\ddot{o}_{1,22}(\varepsilon)} = \frac{4}{7} - \frac{32}{147}\varepsilon + o_2(\varepsilon)$.

Note that the Laurent asymptotic expansion for the expectation $E_{22}(\varepsilon)$ and, in sequel, the Taylor asymptotic expansion for the stationary probability $\pi_2(\varepsilon)$ are invariant with respect to the choice the sequence of states $\langle 1,3\rangle$ or $\langle 3,1\rangle$ for sequential exclusion from the phase space $\mathbb{X}$. This is consistent with the corresponding invariance statements formulated in Theorems 4.1 and 4.2.

The coefficients of the asymptotic expansions $\pi_i(\varepsilon) = c_i[0] + c_i[1]\varepsilon + o_i(\varepsilon), i = 1-3$ given above satisfy relations, $c_1[0] + c_2[0] + c_3[0] = 1$, $c_1[1] + c_2[1] + c_3[1] = 0$.

This is consistent with the corresponding statement in Theorem 4.2.

In the example presented above, we did not trace the explicit formulas for remainders $o_1(\varepsilon), o_2(\varepsilon)$, and $o_3(\varepsilon)$. However, according to the corresponding statement in Theorem 4.2, these remainders are connected by identity, $o_1(\varepsilon) + o_2(\varepsilon) + o_3(\varepsilon) \equiv 0$.

We would like also to explain an unexpected, in some sense, asymptotic behavior of stationary probabilities $\pi_i(\varepsilon)$, in the above example. As a matter of fact, states 1 and 2 are asymptotically absorbing states with non-absorption probabilities of different order, respectively, $O(\varepsilon^2)$ and $O(\varepsilon)$. While, state 3 is a transient asymptotically non-absorbing state. This, seems, should cause convergence of the stationary probability $\pi_1(\varepsilon)$ to 1 and the stationary probabilities $\pi_2(\varepsilon)$ and $\pi_3(\varepsilon)$ to

0 as $\varepsilon \to 0$, with different rates of convergence. This, however, does not take place, and all three probabilities converge to non-zero limits. This is because of the expected sojourn times $e_1(\varepsilon), e_2(\varepsilon)$, and $e_3(\varepsilon)$ have orders, respectively, $O(\varepsilon), O(1)$, and $O(\varepsilon^{-1})$. These expectations compensate absorption effects for states 1, 2, and 3.

6.2.2 Asymptotic Expansions with Explicit Upper Bounds for Remainders

Very useful properties of explicit upper bounds for remainders in asymptotic expansions presented in the book are that they provide "right order" for upper bounds as functions of perturbation parameter ε and that they are asymptotically uniform with respect to perturbation parameter $\varepsilon \to 0$.

At the same time, it should be noted that explicit upper bounds for remainders of asymptotic expansions presented in the book have a conservative "norm-like" character. They are based on the accumulation of absolute values for all components in the corresponding remainders. Absolute values of actual remainders can, in fact, take smaller values due to cancelation effects in the corresponding accumulation processes.

Also, the constants in upper bounds for remainders in the corresponding asymptotic expansions for moments of hitting times and stationary probabilities are sensitive to the dimension of the model determined by the number of states for the underlying semi-Markov process as well as asymptotic singularities of the corresponding matrices of transition characteristics.

These problems do require further in-depth studies, which are beyond the framework of the present book.

Here, we just shortly present an example, by giving explicit upper bounds for remainders in the asymptotic expansions for stationary probability $\pi_3(\varepsilon)$, in the model considered in Subsection 6.2.1.

Let us assume that $\varepsilon_0 = \frac{1}{2}$ and that the remainders in asymptotic expansions for transition probabilities $p_{ij}(\varepsilon)$, given in relation (6.4) satisfy, according condition **F**, the following identities, $o_{11}(\varepsilon) + \varepsilon^2 + \varepsilon^3 + o_{12}(\varepsilon^3)$, $\frac{1}{2}\varepsilon^2 + o_{21}(\varepsilon^2) + o_{22}(\varepsilon) + \frac{1}{2}\varepsilon^2 + o_{23}(\varepsilon^2)$, $o_{31}(\varepsilon) + o_{32}(\varepsilon) \equiv 0$, and admit the explicit upper bounds with parameters $\delta_{ij} = 1, G_{ij} = g, \varepsilon_{ij} = f, j \in \mathbb{Y}_i, i = 1,2,3$, where $g > 0$ and $f = \frac{1}{10}$.

In this case, transition probabilities $p_{ij}(\varepsilon) > 0, \varepsilon \in (0, \varepsilon_0], j \in \mathbb{Y}_i, i \in \mathbb{X}$ and condition **F** holds.

Let us also assume that the remainders in asymptotic expansions for expectations $e_{ij}(\varepsilon)$, given in relation (6.5), also admit the explicit upper bounds with parameters $\dot{\delta}_{ij} = 1, \dot{G}_{ij} = g, \dot{\varepsilon}_{ij} = f = \frac{1}{10}, j \in \mathbb{Y}_i, i = 1,2,3$.

In this case, all expectations $e_{ij}(1,\varepsilon) > 0, \varepsilon \in (0, \varepsilon_0], j \in \mathbb{Y}_i, i \in \mathbb{X}$.

The asymptotic expansion for stationary probability $\pi_3(\varepsilon) = \pi_3(0) + c_3[1]\varepsilon + o_3(\varepsilon) = \frac{2}{7} + \frac{26}{147}\varepsilon + o_3(\varepsilon)$ proposes to use either the approximation $\pi_3(0)$ or the corrected approximation $\pi_3(0) + c_3[1]\varepsilon$ for stationary probability $\pi_3(\varepsilon)$.

The choice of $f = \frac{1}{10}$ is motivated by our intention to keep the correction $c_3[1]\varepsilon$ not exceeding a reasonable limit, not more than $5-6\%$ of $\pi_3(0)$, for $\varepsilon \in (0, f]$.

We compute explicit upper bounds for remainders in asymptotic expansions for the stationary probabilities $\pi_3(\varepsilon)$ using the algorithms given in Theorems 3.5–3.7 and 4.3. We use the non-simplified upper bounds for remainders given in Lemmas 2.1–2.8 and, also, take into account Remark 2.11. We also choose the maximal possible value for parameter α, which guarantees that parameter $\tilde{\varepsilon}'_B \geq f = \frac{1}{10}$ in all cases of application proposition (v) (the division rule) of Lemma 2.6. The corresponding value is $\alpha = \frac{50}{11}$.

The above algorithm yields the explicit upper bounds for the remainder $|o_3(\varepsilon)| \leq G_3(g)\varepsilon^2, \varepsilon \in (0, \frac{1}{10}]$, where $G_3(g) = q_1 + q_2 g + q_3 g^2 \leq 0.4135 + 1.2873g + 0.0028g^2$. Coefficients q_1, q_2, and q_3 are rational numbers, which exact values have been computed, then transformed in the more convenient decimal form and rounded by omitting all digits in the fractional parts beginning from the fifth one and adding 1 to the fourth digit.

Table 6.1 shows the share of correction $c_3[1]\varepsilon$ in the approximation $\pi_3(0) + c_3[1]\varepsilon$ of stationary probability $\pi_3(\varepsilon)$, as well as the value of quotient $G_3(g)\varepsilon^2/c_3[1]\varepsilon$, which corresponds to the upper bound $G_3(g)\varepsilon^2$ for the remainder $o_3(\varepsilon)$, for some typical values of model parameters.

g	ε	$\pi_3(0) \approx$	$c_3[1]\varepsilon \approx$	$G_3(g)\varepsilon^2 \approx$	$c_3[1]\varepsilon/\pi_3(0) \approx$	$G_3(g)\varepsilon^2/c_3[1]\varepsilon \approx$
1	0.1	0.2857	0.0179	0.0170	0.06	0.95
1	0.05	0.2857	0.0089	0.0043	0.03	0.48
1	0.016	0.2857	0.0029	0.0005	0.01	0.17
0.5	0.1	0.2857	0.0179	0.0106	0.06	0.59
0.5	0.05	0.2857	0.0089	0.0027	0.03	0.30
0.5	0.016	0.2857	0.0029	0.0003	0.01	0.10
0.25	0.1	0.2857	0.0179	0.0074	0.06	0.41
0.25	0.05	0.2857	0.0089	0.0019	0.03	0.21
0.25	0.016	0.2857	0.0029	0.0002	0.01	0.02

Table 6.1 Parameters of first order asymptotic expansion with explicit upper bounds for remainder for stationary probability $\pi_3(\varepsilon)$

As one can see, the relations between the terms of asymptotic expansion $\pi_3(0)$, $c_3[1]\varepsilon$ for stationary probability $\pi_3(\varepsilon)$ and the upper bound $G_3(g)\varepsilon^2$ for the corresponding remainder $o_3(\varepsilon)$ are acceptable, in the above example.

6.3 A Survey of Perturbed Stochastic Models

In this section, we give a short survey of perturbed stochastic systems of Markov and semi-Markov types and present typical examples of small perturbation parameters and functionals, for which the corresponding asymptotic perturbation analysis is a natural area for research studies.

6.3.1 Perturbed Queuing Systems

The queuing theory presents a variety of perturbed stochastic models. Small perturbation parameters have often a natural sense for such systems and an asymptotic perturbation analysis plays an important role in their research studies.

These can be different variants of multi-server and multi-repair-unit systems of M/M, M/G, and G/M types, queuing systems with parameters of input flows and service processes depending on a queue, systems with modulated input flows and switching service regimes, systems with several types of customers and several types of servers, with and without reservation of servers, etc. Roles of small perturbation parameters can be played directly by failure intensities for highly reliable systems, or expectations of repairing times for systems with quick repairing, or functions of some perturbation parameter, which represent such characteristics. Times of first hitting into some special subsets of states, interpreted as first failure times or as lifetimes for queuing systems and moments of such hitting times, stationary distributions, and related functionals are usual objects of interest. Asymptotic relations of convergence and rate of convergence types as well as asymptotic expansions, without and with explicit upper bound for remainders, are natural objects for studies of such perturbed systems.

As an example, let us mention a M/M queuing system with N servers functioning independently of each other, with exponentially distributed working and repairing times, failure intensities $\lambda_i^+(\varepsilon), i = 1,\ldots,N$ and repairing intensities $\lambda_i^-(\varepsilon), i = 1,\ldots,N$ depending on some perturbation parameter $\varepsilon \in (0,1]$. In this case, the function of the system is described by Markov chain $\bar{\eta}_\varepsilon(t) = (\eta_1(t),\ldots,\eta_N(t)), t \geq 0$, where $\eta_i(t)$ takes value 1 or 0 if the server i is functioning or not functioning at instant t. the Markov chain $\bar{\eta}_\varepsilon(t)$ has the phase space $\mathbb{X} = \{\bar{i} = (i_1,\ldots,i_N), i_1,\ldots,i_N = 0,1\}$ and transition intensities $\lambda_{\bar{i}\bar{j}}(\varepsilon)$ (here $\bar{j} = (j_1,\ldots,j_N) \in \mathbb{X}$), which take values $\lambda_r^-(\varepsilon)$, for $i_r = 1, j_r = 0, i_k = j_k, k \neq r$; $\lambda_r^+(\varepsilon)$, for $i_r = 0, j_r = 1, i_k = j_k, k \neq r$; and 0, otherwise. In the case, where the failure and repairing intensities can be expanded in asymptotic series with the first non-zero terms of different order, the phase space $\mathbb{X}$ can have an arbitrary asymptotic communicative structure. Objects of interest can be asymptotic expansions for moment of hitting times, stationary and quasi-stationary type distributions, etc.

Another class of queuing systems are M/M, M/G and G/M queuing systems with queue buffers and possible loss of customers.

As an example, let us only mention a simplest M/G queuing system with one server, which has a random service time with distribution $G(\cdot)$, a Poisson input flow of customers with a small input intensity ε, which plays the role of perturbation parameter, and a bounded queue buffer of size N. The service times and the input flow are independent. The object of interest can be moments of loss times (which occur at instants when a customer from the input flow tries to come in the system with full queue buffer), stationary and quasi-stationary type distributions for the corresponding semi-Markov type processes describing dynamics of the number of customers in the system, etc.

There exist very many variants of such systems, with individual or group incoming of customers, characteristics of input flow and services times depending of queue, input flows with Markov modulation, different disciplines of service, vacation periods for empty systems, etc. The role of small perturbation parameters can be played directly by intensities of input flows, or by expectations of service times for systems with quick service, or functions of some perturbation parameter representing these characteristics. Loss times and their moments, stationary distributions of the processes describing the dynamics of the number of customers in such systems, and related stationary functionals are usual objects of interest. Asymptotic relations of convergence and rate of convergence types as well as asymptotic expansions, without and with explicit upper bound for remainders, are natural objects of studies for such perturbed systems.

There exist also many other types of queuing systems with perturbed parameters, for which the asymptotic perturbation analysis is an interesting and important area for studies. Here, we just mention that all queuing systems pointed above also have discrete time analogs. Asymptotic analysis of perturbed discrete time models has its own theoretical and applied values.

We would like to refer here to some books and other works, Korolyuk and Turbin (1976), Borovkov (1984), Kalashnikov (1994), Kovalenko (1994), Kovalenko et al. (1996), Korolyuk and Korolyuk (1999), Whitt (2002), Asmussen (2003), Hassin and Haviv (2003), Bini et al. (2005), Anisimov (2008), Gyllenberg and Silvestrov (2008), and Asmussen and Albrecher (2010), where readers can find descriptions of models, results of their studies, and further related references.

6.3.2 Perturbed Stochastic Networks

There exist a number of types of Markov and semi-Markov stochastic systems, for which their functioning can be interpreted as a passage of some customer (message, production item, etc.) through some stochastic queuing or reliability network and described by some Markov or semi-Markov process. In such models, states in the corresponding phase space are interpreted as servers, the sojourn times in these states as service times (required for transition of messages to some other server, realization of some step in the production process, etc.). The role of small perturbation parameters can be played by probabilities of some transitions, expectations of sojourn times, etc. Hitting times from some input states to some output states representing the global passage times of customers through the network, moments of these random variables as well as stationary distributions for the corresponding Markov and semi-Markov processes are usual objects of interest.

As an example, we can mention Internet based information systems represented by Markov chains associated with the corresponding web links graphs. The famous Google PageRank is based on ranking of web-pages by values of the corresponding stationary probabilities for the Markov chain associated with the corresponding web links graph. The power method is used to approximate the corresponding stationary probabilities π_i by elements of matrix $\mathbf{P}^n$, where $\mathbf{P}$ is the corresponding transition matrix. In order to escape the dangling effects caused by some absorbing pages or groups of pages with no out-links as well as pseudo-dangling effects caused by existence of nearly uncoupled subsets of nodes in the corresponding network graphs, matrix $\mathbf{P}$ is perturbed and approximated by matrices $\mathbf{P}_\varepsilon = (1-\varepsilon)\mathbf{P} + \varepsilon\mathbf{D}$, where $\mathbf{D}$ is some damping stochastic matrix with all positive elements. This is the simplest variant of damping approximations. Analogous problems appear also in more specialized information networks such as P2P networks and DebtRank networks. Mathematics behind the above information network models is very complicated due to the usually huge dimension of matrix $\mathbf{P}$. It is worth to note that, in fact, the damping parameter $0 < \varepsilon < 1$ should be chosen neither too small nor large. In the first case, where ε takes too small values, the damping effect will not work against dangling and pseudo-dangling effects. In the second case, the ranking information accumulated by matrix $\mathbf{P}$ via its stationary distributions will be partly lost, due to the deviation of matrices $\mathbf{P}_\varepsilon$ from matrix $\mathbf{P}$. This actualizes the problem of high order expansions for perturbed stationary probabilities $\pi_i(\varepsilon)$. Such expansions can be objects of interest not only at point 0 but also at points $\dot{\varepsilon} > 0$. In this case, ε should be replaced by a new perturbation parameter $\varepsilon' = \varepsilon - \dot{\varepsilon}$.

In stochastic production networks, some items pass through production, control, reparation, and other type of units and change their quality characteristics during the above multi-stage production process. In such networks, the role of small perturbation parameters can, for example, be played by some small probabilities of failure at some production steps. In communication networks, information items pass through some servers, communication channels, and other units. In such networks, the role of perturbation parameters can, for example, be played by some small probabilities of wrong processing of an information item in some server or small probabilities of wrong transmission of a signal in some communication channel, etc. Natural objects of interest are passage times through the production or communication networks and their moments, probabilities of production of communication success, etc.

Other examples are Markov and semi-Markov networks used in life insurance models, where the corresponding Markov or semi-Markov process can describe a passage of a disabled individual through some stochastic network (with states representing possible disability degrees). The role of perturbation parameters can, for example, be played by small probabilities of transition of an individual to some "rare" state (with a very high disability degree). In non-life insurance models, the corresponding Markov or semi-Markov process can describe a passage of an insured object through some stochastic network (with states representing risk degrees for customers). The role of perturbation parameters can, for example, be played by small probabilities of transition of an insured object to some "rare" states (with high risk degrees). The objects of interest can be hitting times or related accumulated insurance rewards and moments of such functionals. In financial applications, the hitting times for semi-Markov processes can be also interpreted as rewards accumulated up to some hitting terminating times for financial contracts, or default times for Markov-type models describing credit rating dynamics, etc. The role of perturbation parameters can, for example, be played by a small local termination or default probabilities.

An asymptotic analysis plays an important role in research studies of the above perturbed stochastic networks.

We would like to refer here to some books and other works, Kovalenko (1975). Korolyuk and Turbin (1976), Silvestrov (1980), Kovalenko (1994), Kovalenko et al. (1996), Kovalenko et al. (1997), Korolyuk and Korolyuk (1999), Limnios and Oprişan (2001), Langville and Meyer (2006), Janssen and Manca (2007), Andersson and Silvestrov (2008), Barbu and Limnios (2008), Asmussen and Albrecher (2010), Avrachenkov et al. (2013), Engström and Silvestrov (2014, 2016a,b), and Engström (2016), where readers can find descriptions of models, results of their studies, and further related references.

6.3.3 Perturbed Bio-Stochastic Systems

A very large class of perturbed stochastic models constitute biological type systems, in particular models of population dynamics, epidemic models, and models of mathematical genetics. Processes of birth-death-type play here a very important role.

Initial characteristics for Markov models of population dynamics can be rates (intensities) of birth, death, immigration, and emigration. The role of perturbation parameter can, for example, be played by a small immigration rate. In such finite models, the population faces subsequent extinction and recolonization events. Another perturbation scenario is, where the role of small perturbation parameter is played by a small birth rate. In both cases, the Markov chain representing the dynamics of population size has the asymptotically absorbing state 0.

In epidemic models for finite populations the initial characteristics can be contamination, individual recovering rates (intensities) as well as contamination rate via external contacts of individuals with infected ones outside of the population. The latter rate can, for example, play the role of small perturbation parameter. In more complex cases, all above intensities may depend on a small perturbation parameter and converge to some limiting values when the perturbation parameter tends to zero. In such models, the epidemic periodically disappears and appears again in the population. The Markov chain, which represents the dynamics of epidemic size, has the asymptotically absorbing state 0.

The objects of interest for the above perturbed population dynamics and epidemic models can be moments of extinction times, stationary distributions, and some of their modifications such as conditional stationary distributions. The asymptotic perturbation analysis allows one to investigate sensitivity of functionals of interest with respect to small changes of model parameters or to get more advanced information about their asymptotic behavior in the form of asymptotic expansions.

One of the basic models of population genetics relates to one-sex finite population of a size N. Individuals from this population carry two copies of a certain gene, which can exist in two variants (alleles), say, of types A and B. Dynamics of the number of individuals with the above gene of type A is described by a Markov chain of birth-death type with intensities of $\lambda_i^\pm$ of ± 1 jumps in a non-end states $0 < i < N$. The classical case is where intensities $\lambda_i^+ = \lambda^+ \frac{i}{N}$ and $\lambda_i^- = \lambda^- \frac{N-i}{N}$, for $0 < i < N$. These end states, 0 and N, are absorbing in a non-perturbed model. A small perturbation parameter $\varepsilon > 0$ can be implemented in the model in such a way that some small positive intensities, $\lambda_0^+(\varepsilon) = \lambda_0^+ \varepsilon$ of $+1$ jump and $\lambda_N^-(\varepsilon) = \lambda^- \varepsilon$ of -1 jump, are assigned, respectively, for states 0 and N. Intensities of ± 1 jumps for non-end states also can be functions of this small perturbation mutation parameter. For example, these intensities can be $\lambda_i^+(\varepsilon) = \lambda^+(1-\varepsilon)\frac{i}{N} + \lambda^- \varepsilon \frac{N-i}{N}$ and $\lambda_i^-(\varepsilon) = \lambda^+ \varepsilon \frac{i}{N} + \lambda^-(1-\varepsilon)\frac{N-i}{N}$, for $0 < i < N$. In such models, the population can either have a mixed gene configuration, or have one of the two extremal gene configurations, where all individuals have either gene of type A or gene of type B. The objects of interest for such perturbed models can be the first absorption (hitting to state 0 or N) times and their moments, probabilities of first hitting one of the two end states, stationary and conditional quasi-stationary distributions, etc. Asymptotic perturbation analysis of such models is one of the important applied problems.

There exist a large number of models of biological type generalizing those described above. For example, these are meta-population models with several sub-populations possessing birth-death-type dynamics. In such model birth and death intensities in sub-populations and migration intensities between sub-populations can be functions of some perturbation parameter. The whole population can be asymptotically split into several non-communicated parts.

Also, models described above have discrete time analogs, semi-Markov, and many other variants of generalization. Nonlinearly perturbed epidemic and population dynamics models and models of population genetics with different types, e.g., sex, age, health status, etc., of individuals in the population, forms of interactions between individuals, environmental modulation, complex spatial structure, etc., also make interesting objects for asymptotic perturbation analysis.

We would like to refer here to some books and other works, Gyllenberg and Hanski (1992), Hanski and Gilpin (1997), Hanski (1999), Daley and Gani (1999), Lande et al. (2003), Durrett (2008), Gyllenberg and Silvestrov (2008), Nåsell (2011), Allen and Tarnita (2014), and Silvestrov et al. (2016), where readers can find descriptions of models, results of their studies, and further related references.

6.3.4 Other Classes of Perturbed Stochastic Systems

There are other classes of applied perturbed models of stochastic systems, for which asymptotic perturbation analysis plays an important role, for example, Markov decision models, control systems, etc.

We would like to refer here to some books and other works, Abbad and Filar (1995), Avrachenkov et al. (2002), Yin and Zhang (2005, 2013), Borkar et al. (2012), Avrachenkov et al. (2013), and Obzherin and Boyko (2015) where readers can find descriptions of models, results of their studies, and further related references.

Additional references to works connected with asymptotic perturbation analysis for perturbed stochastic systems and processes are also given in the bibliographical remarks given in Appendix A.

Appendix A
Methodological and Bibliographical Remarks

In this appendix, we present some methodological and bibliographical remarks connected with the content of the book and comment new results presented in the book and new directions for future research.

A.1 Methodological Remarks

Semi-Markov processes were introduced independently by Lévy (1954), Takács (1954), and Smith (1955) as a natural generalization of discrete and continuous time Markov chains. These processes are a very flexible and effective tool for analysis of queuing, reliability and control systems, biological systems, information networks, financial and insurance processes and many other stochastic systems and processes.

We would like to refer here to the books and survey papers devoted to semi-Markov and related processes that are Korolyuk et al. (1974), Korolyuk and Turbin (1976), Silvestrov (1980), Volker (1980), Korolyuk and Swishchuk (1995), Limnios and Oprişan (2001), Janssen and Manca (2006, 2007), Howard (2007), Barbu and Limnios (2008), Gyllenberg and Silvestrov (2008), Harlamov (2008), Grabski (2015), and Obzherin and Boyko (2015).

The model of perturbed Markov chains and semi-Markov processes, in particular, in the most difficult cases of perturbed processes with absorption and so-called singularly perturbed processes, attracted attention of researchers in the end of fifties of the twenties century. The first works related to asymptotical problems for the above models are Meshalkin (1958), Simon and Ando (1961), Blackwell (1962), Hanen (1963), Seneta (1967, 1968a,b), Schweitzer (1968), Korolyuk (1969), Miller and Veinott (1969), and Veinott (1969).

D. Silvestrov, S. Silvestrov, *Nonlinearly Perturbed Semi-Markov Processes*,
SpringerBriefs in Probability and Mathematical Statistics,
DOI 10.1007/978-3-319-60988-1

The methods used for construction of asymptotic expansions for moments of hitting times, stationary distributions and related functionals can be split into three groups.

The first and the most widely used methods are based on analysis of generalized matrix and operator inverses of resolvent type for transition matrices and operators for singularly perturbed Markov chains and semi-Markov processes. Mainly models with linear, polynomial and analytic perturbations of transition characteristics have been objects of studies. We refer here to works by Schweitzer (1968, 1987), Korolyuk and Turbin (1970, 1976, 1978), Turbin (1972), Poliščuk and Turbin (1973), Pervozvanskiĭand Smirnov (1974), Courtois and Louchard (1976), Courtois (1977), Latouche and Louchard (1978), Kokotović et al. (1980), Korolyuk et al. (1981), Phillips and Kokotović (1981), Delebecque (1983), Stewart (1993b, 1998, 2001), Abadov (1984), Silvestrov and Abadov (1984, 1991, 1993), Kartashov (1985, 1996), Haviv (1986), Korolyuk (1989), Stewart and Sun (1990), Haviv et al. (1992), Haviv and Ritov (1993), Schweitzer and Stewart (1993), Avrachenkov (1999, 2000), Avrachenkov and Lasserre (1999), Korolyuk and Korolyuk (1999), Yin et al. (2001), Avrachenkov and Haviv (2003, 2004), Craven (2003), Yin and Zhang (2003, 2005, 2013), Bini et al. (2005), Koroliuk and Limnios (2005), Aliev and Abadov (2007), Avrachenkov et al. (2013), and Avrachenkov et al. (2016).

Aggregation/disaggregation methods based on various modification of Gauss elimination method and space screening procedures for perturbed Markov chains have been employed for approximation of stationary distributions for Markov chains in works by Coderch et al. (1983), Delebecque (1983), Gaĭtsgori and Pervozvanskiy (1983), Chatelin and Miranker (1984), Courtois and Semal (1984), Seneta (1984), Cao and Stewart (1985), Schweitzer et al. (1985), Vantilborgh (1985), Schweitzer and Kindle (1986), Feinberg and Chiu (1987), Haviv (1987, 1992, 1999), Sumita and Reiders (1988), Meyer (1989), Schweitzer (1991), Stewart and Zhang (1991), Stewart (1993a), Kim and Smith (1995), Marek and Pultarová (2006), Marek et al. (2009), and Avrachenkov et al. (2013).

Alternatively, methods based on regenerative properties of Markov chains and semi-Markov processes, in particular, relations which link stationary probabilities and expectations of return times have been used for getting approximations for expectations of hitting times and stationary distributions in works by Grassman et al. (1985), Hassin and Haviv (1992), and Hunter (2006, 2014). Also, the above-mentioned relations and methods based on asymptotic expansions for nonlinearly perturbed regenerative processes developed in works by Silvestrov (1995, 2007, 2010), Englund and Silvestrov (1997), Gyllenberg and Silvestrov (1998, 1999, 2000, 2008), Englund (2000, 2001), Ni et al. (2008), Ni (2010a,b, 2011, 2012, 2014), Petersson (2013, 2014, 2016a,b,c), and Silvestrov and Petersson (2013) have been used for getting asymptotic expansions for moments of hitting times, stationary and quasi-stationary distributions for nonlinearly perturbed Markov chains and semi-Markov processes with absorption.

We would like especially to mention books including material on perturbed Markov chains, semi-Markov processes and related problems. These are Seneta (1973, 2006), Silvestrov (1974, 2004), Kovalenko (1975), Korolyuk and Turbin

(1976, 1978), Courtois (1977), Kalashnikov (1978, 1997), Anisimov (1988, 2008), Korolyuk (1989), Stewart and Sun (1990), Korolyuk and Swishchuk (1995), Kartashov (1996), Kovalenko et al. (1996), Borovkov (1998), Stewart (1998, 2001), Korolyuk and Korolyuk (1999), Koroliuk and Limnios (2005), Yin and Zhang (2005, 2013), Gyllenberg and Silvestrov (2008), Meyn and Tweedie (2009), Avrachenkov et al. (2013), and Obzherin and Boyko (2015).

Also, we would like to mention the recent paper by Silvestrov and Silvestrov (2016a), where a comprehensive bibliography of works in the related areas can be found.

A.2 New Results Presented in the Book

In the present book, we combine methods based on stochastic aggregation/disaggregation approach with methods based on asymptotic expansions for perturbed regenerative processes applied to perturbed semi-Markov processes.

In the above-mentioned works based on stochastic aggregation/disaggregation approach, space screening procedures for discrete time Markov chains are used. A Markov chain with a reduced phase space is constructed from the initial one as the sequence of its states at sequential moments of hitting into the reduced phase space. Such screening procedure preserves ratios of hitting frequencies for states from the reduced phase space and, thus, the ratios of stationary probabilities are the same for the initial and the reduced Markov chains. This implies that the stationary probabilities for the reduced Markov chain coincide with the corresponding stationary probabilities for the initial Markov chain up to the change of the corresponding normalizing factors.

We use another more complex type of time-space screening procedures for semi-Markov processes. In this case, a semi-Markov process with a reduced phase space is constructed from the initial one as the sequence of its states at sequential moments of hitting into the reduced phase space, and times between sequential jumps of the reduced semi-Markov process are times between sequential hitting of the reduced space by the initial semi-Markov process. Such screening procedure preserves hitting times between states from the reduced phase space, i.e., these times and, thus, their moments are the same for the initial and the reduced semi-Markov processes.

It should be mentioned that the semi-Markov setting is an adequate and, moreover, necessary element of the proposed method. Even in the case, where the initial process is a discrete or continuous time Markov chain, the time-space screening procedure of phase space reduction results in a semi-Markov process, since times between sequential hitting of the reduced space by the initial process have distributions which can differ from geometrical or exponential ones.

We also formulate perturbation conditions in terms of Taylor asymptotic expansions for transition probabilities of the corresponding embedded Markov chains and Laurent asymptotic expansions for power moments of transition times for perturbed semi-Markov processes. The remainders in these expansions and, thus, the transition characteristics of perturbed semi-Markov processes can be non-analytic functions of perturbation parameter. This makes difference with the results for models with linear, polynomial, and analytic perturbations.

The use of Laurent asymptotic expansions for moments of transition times for perturbed semi-Markov processes is also a necessary element of the method. Indeed, even in the case, where moments of transition times for all states of the initial semi-Markov process are asymptotically bounded and represented by Taylor asymptotic expansions, the exclusion of an asymptotically absorbing state from the initial phase space can cause appearance of states with asymptotically unbounded moments of transition times represented by Laurent asymptotic expansions, for the corresponding reduced semi-Markov processes.

An important novelty of our results also is that we have got asymptotic expansions with remainders given not only in the standard form of $o(\cdot)$, but also, in a more advanced form, with explicit power-type upper bounds for remainders, asymptotically uniform with respect to a perturbation parameter. Such asymptotic expansions for nonlinearly perturbed semi-Markov processes were not known before.

Semi-Markov processes are a natural generalization of Markov chains, important theoretically and essentially extending applications of Markov-type models. The asymptotic results presented in the book are a good illustration for this statement. In particular, they automatically yield analogous asymptotic results for nonlinearly perturbed discrete and continuous time Markov chains.

In the present book, we continue the line of research approach to asymptotical analysis of nonlinearly perturbed semi-Markov processes developed in the book by Gyllenberg and Silvestrov (2008). In this book, generalized matrix inverses have been used for getting Taylor asymptotic expansions for power and mixed power-exponential moments of hitting times, stationary and quasi-stationary distributions, and related functionals for singularly perturbed semi-Markov processes with absorption, with a simple asymptotic structure of the phase space consisting of one closed communicative class of states plus possibly a class of transient states. This method, however, does not work well for the more complex model of singularly perturbed semi-Markov processes with a more complex asymptotic structure of phase space consisting of several closed communicative classes of states plus possibly a class of transient states. In this case, moments of hitting times can be asymptotically unbounded functions of perturbation parameter due to the presence of asymptotically absorbing states or subsets of states. Their asymptotic analysis, with the use of the generalized matrix inverses, becomes rather intricate.

In the present book, we develop a new method based on recurrent reduction of phase spaces combined with the use of techniques of more general Laurent asymptotic expansions. This method has serious advantages.

First, this method made it possible to get asymptotic expansions for power moments of hitting times, stationary distributions, and related functionals for singularly perturbed semi-Markov processes with an arbitrary asymptotic structure of the phase space, including the general case, where the phase space can be asymptotically split into several closed classes of communicative states and, possibly, a class of transient states.

Second, the proposed recurrent algorithms are much more computationally effective for models with non-singular and singular perturbation. It has a clear recurrent form well prepared, for example, for effective programming. These algorithms also possess good pivotal properties, which make it possible to postpone exclusion of nearly absorbing states. This makes computation of transition characteristics for reduced semi-Markov processes numerically more stable.

Third, the proposed algorithms let us get not only asymptotic expansions with remainders given in the standard form of $o(\cdot)$, but also new more advanced asymptotic expansions with explicit power upper bounds for remainders.

Fourth, the proposed method lets us get asymptotic expansions for nonlinearly perturbed birth-death-type semi-Markov processes with finite phase spaces under much more general perturbation conditions than in the above book and, also, in two forms, without and with explicit upper bounds for remainders.

In the present book, we restrict consideration to power moments of hitting times, stationary distributions, and related functionals. However, we are quite sure that the method based on sequential reduction of phase space and the corresponding recurrent algorithms will make it possible to get asymptotic expansions for mixed power-exponential moments and quasi-stationary distributions for singularly perturbed semi-Markov processes.

Finally, we would like to comment on a calculus of Laurent asymptotic expansions thoroughly presented in the book. Operational rules for such expansions, with remainders given in the standard form $o(\cdot)$ should be possibly considered as generally known, while operational rules for Laurent asymptotic expansions with explicit power-type upper bounds have some novelty aspects.

The book is partly based on results presented in the recent papers by Silvestrov and Silvestrov (2016a,b). In Chapter 2, the operational rules for Laurent asymptotic expansions given in these papers are presented in extended and improved form. In particular, simplified explicit upper bounds for remainders are added. In Chapter 3, the results of the above papers about Laurent asymptotic expansions of expectations of hitting times for perturbed semi-Markov processes are generalized to power moments of higher order. Chapter 4 presents an improved and extended version of results of the above papers concerning asymptotic expansions for stationary distributions of perturbed semi-Markov processes as well as new results about asymptotic expansions for conditional quasi-stationary distributions and other stationary limits. In Chapter 5, we generalize some results presented in the recent paper by Silvestrov et al. (2016). Sections 5.2 and 5.3 contain an improved version of the corresponding results on asymptotic expansions for stationary and quasi-stationary distributions of perturbed semi-Markov processes of birth-death type, in particular, new results about asymptotic expansions with explicit upper bounds for remainders. Chapter 6 presents an extended variant of numerical examples from the paper by Silvestrov and Silvestrov (2016b).

A.3 New Problems

In the present book, we consider the model, where the pre-limiting perturbed semi-Markov processes have a phase space which is one class of communicative states asymptotically splitting into one or several classes of communicative states and possibly a class of transient states. However, the method of sequential reduction of the phase space can also be applied to models, where the pre-limiting semi-Markov processes also have a phase space, which consists of several classes of communicative states and a class of transient states.

The problems also requiring additional studies are concerned with aggregation of steps in the time-space screening procedures for semi-Markov processes, tracing pivotal orders for different groups of states as well as getting explicit formulas for coefficients of expansions and parameters of upper bounds for remainders in the corresponding asymptotic expansions for moments of hitting times as well as for stationary and conditional quasi-stationary distributions. Such formulas can be obtained, for example, for birth-death-type semi-Markov processes, for which the proposed algorithms of phase space reduction preserve the birth-death structure for reduced semi-Markov processes. Some initial results in this direction are presented in the recent paper by Silvestrov et al. (2016) and Chapter 5 of the present book, where the phase space of pre-limiting birth-death-type semi-Markov processes asymptotically splits into one communicative class of transient states and one or two absorbing end states. The above paper also contains results on asymptotic expansions for stationary and conditional quasi-stationary distributions for models of population dynamics, epidemic models, and models of population genetics. We are sure that these results can be generalized to the so-called star-birth-death-type semi-Markov processes, which have phase spaces asymptotically splitting into several closed classes of states with the above described structure and small migration rates between these classes.

The open problem also is that of getting asymptotic expansions with alternative forms for remainders. In particular, this relates to upper bounds possessing algebraic properties making them invariant with respect to the choice of sequences of states sequentially excluded from a phase space.

We would like again to mention interesting problems connected with getting asymptotic expansions for mixed power-exponential moment of hitting times and quasi-stationary distributions (commented in Subsection 4.3.1) for nonlinearly perturbed semi-Markov processes.

An interesting direction for future studies is connected with non-polynomial asymptotic expansions for nonlinearly and singularly perturbed semi-Markov processes. Some related results can be found in papers by Ni et al. (2008) and Ni (2010a,b, 2011, 2012, 2014).

A very prospective new direction for future studies is also connected with problems of getting asymptotic expansions for nonlinearly perturbed semi-Markov processes with multivariate perturbation parameters.

The results presented in the book can be also generalized to more general models, in particular, nonlinearly perturbed stochastic processes with semi-Markov modulation.

Experimental numerical studies, based on the effective computer program realization of recurrent algorithms based on the method of sequential phase space reduction, are also important for the future research in the area.

Finally, we would like to mention an unbounded area of applications to perturbed queuing, reliability and control systems, information, communication and production and other stochastic networks, epidemic models and models of population dynamics and mathematical genetics.

References

Abadov, Z.A.: Asymptotical expansions with explicit estimation of constants for exponential moments of sums of random variables defined on a Markov chain and their applications to limit theorems for first hitting times. Candidate of Science Dissertation, Kiev State University (1984)

Abbad, M., Filar, J.A.: Algorithms for singularly perturbed Markov control problems: a survey. In: Leondes, C.T. (ed.) Techniques in Discrete-Time Stochastic Control Systems. Control and Dynamic Systems, vol. 73, pp. 257–289. Academic Press, New York (1995)

Aliev, S.A., Abadov, Z.A.: A new approach for studying asymptotics of exponential moments of sums of random variables connected in Markov chain. I. Trans. Inst. Math. Mech. Nat. Acad, Sci. Azerb. **27**(7), 17–32 (2007)

Allen, B., Tarnita, C.E.: Measures of success in a class of evolutionary models with fixed population size and structure. J. Math. Biol. **68**, 109–143 (2014)

Andersson, F., Silvestrov, S.: The mathematics of Internet search engines. Acta Appl. Math. **104**, 211–242 (2008)

Anisimov, V.V.: Random Processes with Discrete Components, 183 pp. Vysshaya Shkola and Izdatel'stvo Kievskogo Universiteta, Kiev (1988)

Anisimov, V.V.: Switching Processes in Queueing Models. Applied Stochastic Methods Series, 345 pp. ISTE, London and Wiley, Hoboken, NJ (2008)

Asmussen, S.: Applied Probability and Queues. 2nd edn. Applications of Mathematics, vol. 51. Stochastic Modelling and Applied Probability, xii+438 pp. Springer, New York (2003)

Asmussen, S., Albrecher, H.: Ruin Probabilities, 2nd edn. Advanced Series on Statistical Science & Applied Probability, vol. 14, xviii+602 pp. World Scientific, Singapore (2010)

Avrachenkov, K.E.: Analytic perturbation theory and its applications. Ph.D. Thesis, University of South Australia (1999)

Avrachenkov, K.E.: Singularly perturbed finite Markov chains with general ergodic structure. In: Boel, R., Stremersch, G. (eds.) Discrete Event Systems. Analysis and Control. Kluwer International Series in Engineering and Computer Science, vol. 569, pp. 429–432. Kluwer, Boston (2000)

Avrachenkov, K.E., Haviv, M.: Perturbation of null spaces with application to the eigenvalue problem and generalized inverses. Linear Algebra Appl. **369**, 1–25 (2003)

Avrachenkov, K.E., Haviv, M.: The first Laurent series coefficients for singularly perturbed stochastic matrices. Linear Algebra Appl. **386**, 243–259 (2004)

Avrachenkov, K.E., Lasserre, J.B.: The fundamental matrix of singularly perturbed Markov chains. Adv. Appl. Probab. **31**(3), 679–697 (1999)

D. Silvestrov, S. Silvestrov, *Nonlinearly Perturbed Semi-Markov Processes*,
SpringerBriefs in Probability and Mathematical Statistics,
DOI 10.1007/978-3-319-60988-1

Avrachenkov, K.E., Filar, J., Haviv, M.: Singular perturbations of Markov chains and decision processes. In: Feinberg, E.A., Shwartz, A. (eds.) Handbook of Markov Decision Processes. Methods and Applications. International Series in Operations Research & Management Science, vol. 40, pp. 113–150. Kluwer, Boston (2002)

Avrachenkov, K.E., Filar, J.A., Howlett, P.G.: Analytic Perturbation Theory and Its Applications, xii+372 pp. SIAM, Philadelphia, PA (2013)

Avrachenkov, K., Eshragh, A., Filar, J.A.: On transition matrices of Markov chains corresponding to Hamiltonian cycles. Ann. Oper. Res. **243**(1–2), 19–35 (2016)

Barbu, V.S., Limnios, N.: Semi-Markov Chains and Hidden Semi-Markov Models Toward Applications. Their Use in Reliability and DNA Analysis. Lecture Notes in Statistics, vol. 191, xiv+224 pp. Springer, New York (2008)

Bini, D.A., Latouche, G., Meini, B.: Numerical Methods for Structured Markov Chains. Numerical Mathematics and Scientific Computation. Oxford Science Publications, xii+327 pp. Oxford University Press, New York (2005)

Blackwell, D.: Discrete dynamic programming. Ann. Math. Stat. **33**, 719–726 (1962)

Borkar, V.S., Ejov, V., Filar, J.A., Nguyen, G.T.: Hamiltonian Cycle Problem and Markov Chains. International Series in Operations Research & Management Science, vol. 171, xiv+201 pp. Springer, New York (2012)

Borovkov, A.A.: Asymptotic Methods in Queuing Theory. Wiley Series in Probability and Mathematical Statistics: Applied Probability and Statistics, xi+292 pp. Wiley, Chichester (1984) (English translation of Asymptotic Methods in Queuing Theory. Probability Theory and Mathematical Statistics, 382 pp. Nauka, Moscow (1980))

Borovkov, A.A.: Ergodicity and Stability of Stochastic Processes. Wiley Series in Probability and Statistics, vol. 314, xxiv+585 pp. Wiley, Chichester (1998) (Translation from the 1994 Russian original)

Cao, W.L., Stewart, W.J.: Iterative aggregation/disaggregation techniques for nearly uncoupled Markov chains. J. Assoc. Comput. Mach. **32**, 702–719 (1985)

Chatelin, F., Miranker, W.L.: Aggregation/disaggregation for eigenvalue problems. SIAM J. Numer. Anal. **21**(3), 567–582 (1984)

Coderch, M., Willsky, A.S., Sastry, S.S., Castañon, D.A.: Hierarchical aggregation of singularly perturbed finite state Markov processes. Stochastics **8**, 259–289 (1983)

Collet, P., Martínez, S., San Martín, J.: Quasi-Stationary Distributions. Markov Chains, Diffusions and Dynamical Systems. Probability and its Applications, xvi+280 pp. Springer, Heidelberg (2013)

Courtois, P.J.: Decomposability: Queueing and Computer System Applications. ACM Monograph Series, xiii+201 pp. Academic Press, New York (1977)

Courtois, P.J., Louchard, G.: Approximation of eigen characteristics in nearly-completely decomposable stochastic systems. Stoch. Process. Appl. **4**, 283–296 (1976)

Courtois, P.J., Semal, P.: Block decomposition and iteration in stochastic matrices. Philips J. Res. **39**(4–5), 178–194 (1984)

Craven, B.D.: Perturbed Markov processes. Stoch. Models **19**(2), 269–285 (2003)

Daley, D.J., Gani, J.: Epidemic Modelling: An Introduction. Cambridge Studies in Mathematical Biology, vol. 15, xii+213 pp. Cambridge University Press, Cambridge (1999)

Delebecque, F.: A reduction process for perturbed Markov chains. SIAM J. Appl. Math. **43**, 325–350 (1983)

Durrett, R.: Probability Models for DNA Sequence Evolution, xii+431 pp. Springer, New York (2008) (2nd revised edition of Probability Models for DNA Sequence Evolution, viii+240 pp. Springer, New York (2002))

Englund, E.: Nonlinearly perturbed renewal equations with applications to a random walk. In: Silvestrov, D., Yadrenko, M., Olenko A., Zinchenko, N. (eds.) Proceedings of the Third International School on Applied Statistics, Financial and Actuarial Mathematics, Feodosiya, 2000. Theory Stoch. Process. **6**(22)(3–4), pp. 33–60 (2000)

Englund, E.: Nonlinearly perturbed renewal equations with applications. Doctoral Dissertation, Umeå University (2001)

Englund, E., Silvestrov, D.S.: Mixed large deviation and ergodic theorems for regenerative processes with discrete time. In: Jagers, P., Kulldorff, G., Portenko, N., Silvestrov, D. (eds.) Proceedings of the Second Scandinavian–Ukrainian Conference in Mathematical Statistics, Vol. I, Umeå, 1997. Theory Stoch. Process. **3**(19)(1–2), pp. 164–176 (1997)

Engström, C.: PageRank in evolving networks and applications of graphs in natural language processing and biology. Doctoral Dissertation, 217, Mälardalen University, Västerås (2016)

Engström, C., Silvestrov, S.: Generalisation of the damping factor in PageRank for weighted networks. In: Silvestrov, D., Martin-Löf, A. (eds.) Modern Problems in Insurance Mathematics, chap. 19. EAA Series, pp. 313–334. Springer, Cham (2014)

Engström, C., Silvestrov, S.: PageRank, a look at small changes in a line of nodes and the complete graph. In: Silvestrov, S., Rančić, M. (eds.) Engineering Mathematics II. Algebraic, Stochastic and Analysis Structures for Networks, Data Classification and Optimization, chap. 11. Springer Proceedings in Mathematics & Statistics, vol. 179, pp. 223–248. Springer, Cham (2016a)

Engström, C., Silvestrov, S.: PageRank, connecting a line of nodes with a complete graph. In: Silvestrov, S., Rančić, M. (eds.) Engineering Mathematics II. Algebraic, Stochastic and Analysis Structures for Networks, Data Classification and Optimization, chap. 12. Springer Proceedings in Mathematics & Statistics, vol. 179, pp. 249–274. Springer, Cham (2016b)

Feinberg, B.N., Chiu, S.S.: A method to calculate steady-state distributions of large Markov chains by aggregating states. Oper. Res. **35**(2), 282–290 (1987)

Feller, W.: An Introduction to Probability Theory and Its Applications. Vol. I, xviii+509 pp. Wiley, New York (1968) (3rd edition of An Introduction to Probability Theory and Its Applications. Vol. I, xii+419 pp. Wiley, New York (1950))

Gaĭtsgori, V.G., Pervozvanskiy, A.A.: Decomposition and aggregation in problems with a small parameter. Izv. Akad. Nauk SSSR. Tekhn. Kibernet. (1) 33–46 (1983) (English translation in Eng. Cybernetics **2**(1), 26–38)

Grabski, F.: Semi-Markov Processes: Applications in System Reliability and Maintenance, xiv+255 pp. Elsevier, Amsterdam (2015)

Grassman, W.K., Taksar, M.I., Heyman, D.P.: Regenerative analysis and steady state distributions for Markov chains. Oper. Res. **33**, 1107–1116 (1985)

Gyllenberg, M., Hanski, I.: Single-species metapopulation dynamics: a structured model. Theor. Popul. Biol. **42**(1), 35–61 (1992)

Gyllenberg, M., Silvestrov, D.S.: Quasi-stationary phenomena in semi-Markov models. In: Janssen, J., Limnios, N. (eds.) Proceedings of the Second International Symposium on Semi-Markov Models: Theory and Applications, Compiègne, 1998, pp. 87–93 (1998)

Gyllenberg, M., Silvestrov, D.S.: Quasi-stationary phenomena for semi-Markov processes. In: Janssen, J., Limnios, N. (eds.) Semi-Markov Models and Applications, pp. 33–60. Kluwer, Dordrecht (1999)

Gyllenberg, M., Silvestrov, D.S.: Nonlinearly perturbed regenerative processes and pseudo-stationary phenomena for stochastic systems. Stoch. Process. Appl. **86**, 1–27 (2000)

Gyllenberg, M., Silvestrov, D.S.: Quasi-Stationary Phenomena in Nonlinearly Perturbed Stochastic Systems. De Gruyter Expositions in Mathematics, vol. 44, ix+579 pp. Walter de Gruyter, Berlin (2008)

Hanen, A.: Théorèmes limites pour une suite de chaînes de Markov. Ann. Inst. H. Poincaré **18**, 197–301 (1963)

Hanski, I.: Metapopulation Ecology. Oxford Series in Ecology and Evolution, 324 pp. Oxford University Press, Oxford (1999)

Hanski, I., Gilpin, M.E.: Metapopulation Biology: Ecology, Genetics, and Evolution, 512 pp. Academic Press, London (1997)

Harlamov, B.: Continuous Semi-Markov Processes. Applied Stochastic Methods Series, 375 pp. ISTE, London and Wiley, Hoboken, NJ (2008)

Hassin, R., Haviv, M.: Mean passage times and nearly uncoupled Markov chains. SIAM J. Discrete Math. **5**, 386–397 (1992)

Hassin, R., Haviv, M.: To Queue or not to Queue: Equilibrium Behavior in Queueing Systems. International Series in Operations Research & Management Science, vol. 59, xii+191 pp. Kluwer, Boston (2003)

Haviv, M.: An approximation to the stationary distribution of a nearly completely decomposable Markov chain and its error analysis. SIAM J. Algebraic Discrete Methods **7**(4), 589–593 (1986)

Haviv, M.: Aggregation/disaggregation methods for computing the stationary distribution of a Markov chain. SIAM J. Numer. Anal. **24**(4), 952–966 (1987)

Haviv, M.: An aggregation/disaggregation algorithm for computing the stationary distribution of a large Markov chain. Commun. Stat. Stoch. Models **8**(3), 565–575 (1992)

Haviv, M.: On censored Markov chains, best augmentations and aggregation/disaggregation procedures. Aggregation and disaggregation in operations research. Comput. Oper. Res. **26**(10–11), 1125–1132 (1999)

Haviv, M., Ritov, Y.: On series expansions and stochastic matrices. SIAM J. Matrix Anal. Appl. **14**(3), 670–676 (1993)

Haviv, M., Ritov, Y., Rothblum, U.G.: Taylor expansions of eigenvalues of perturbed matrices with applications to spectral radii of nonnegative matrices. Linear Algebra Appl. **168**, 159–188 (1992)

Hörmander, L.: An Introduction to Complex Analysis in Several Variables, 3rd edn. North-Holland Mathematical Library, vol. 7, xii+254 pp. North-Holland, Amsterdam (1990) (3rd edition of An Introduction to Complex Analysis in Several Variables, x+208 pp. D. Van Nostrand, Princeton (1966))

Howard, R.A.: Dynamic Probabilistic Systems, Volume II: Semi-Markov and Decision Processes. Dover Books on Mathematics, 576 pp. Dover, New York (2007) (Reprint of Wiley 1971 edition)

Hunter, J.J.: Mixing times with applications to perturbed Markov chains. Linear Algebra Appl. **417**(1), 108–123 (2006)

Hunter, J.J.: Generalized inverses of Markovian kernels in terms of properties of the Markov chain. Linear Algebra Appl. **447**, 38–55 (2014)

Janssen, J., Manca, R.: Applied Semi-Markov Processes, xii+309 pp.. Springer, New York (2006)

Janssen, J., Manca, R.: Semi-Markov Risk Models for Finance, Insurance and Reliability, xvii+429 pp. Springer, New York (2007)

Kalashnikov, V.V.: Qualitative Analysis of the Behaviour of Complex Systems by the Method of Test Functions. Series in Theory and Methods of Systems Analysis, 247 pp. Nauka, Moscow (1978)

Kalashnikov, V.V.: Mathematical Methods in Queuing Theory. Mathematics and Its Applications, vol. 271, x+377 pp. Kluwer, Dordrecht (1994)

Kalashnikov, V.V.: Geometric Sums: Bounds for Rare Events with Applications. Mathematics and Its Applications, vol. 413, ix+265 pp. Kluwer, Dordrecht (1997)

Karlin, S., Taylor, H.M.: A First Course in Stochastic Processes, 2nd edn., xvii+557 pp. Academic Press, New York (1975)

Kartashov, N.V.: Asymptotic expansions and inequalities in stability theorems for general Markov chains under relatively bounded perturbations. In: Stability Problems for Stochastic Models, Varna, pp. 75–85. VNIISI, Moscow (1985) (English translation in J. Sov. Math. **40**(4), 509–518)

Kartashov, M.V.: Strong Stable Markov Chains, 138 pp. VSP, Utrecht and TBiMC, Kiev (1996)

Kim, D.S., Smith, R.L.: An exact aggregation/disaggregation algorithm for large scale Markov chains. Nav. Res. Logist. **42**(7), 1115–1128 (1995)

Kokotović, P.V., Phillips, R.G., Javid, S.H.: Singular perturbation modeling of Markov processes. In: Bensoussan, A., Lions, J.L. (eds.) Analysis and Optimization of Systems: Proceedings of the Fourth International Conference on Analysis and Optimization. Lecture Notes in Control and Information Science, vol. 28, pp. 3–15. Springer, Berlin (1980)

Korolyuk, V.S.: On asymptotical estimate for time of a semi-Markov process being in the set of states. Ukr. Mat. Zh. **21**, 842–845 (1969) (English translation in Ukr. Math. J. **21**, 705–710)

Korolyuk, V.S.: Stochastic Models of Systems, 208 pp. Naukova Dumka, Kiev (1989)

Korolyuk, V.S., Korolyuk, V.V.: Stochastic Models of Systems. Mathematics and Its Applications, vol. 469, xii+185 pp. Kluwer, Dordrecht (1999)

Koroliuk, V.S., Limnios, N.: Stochastic Systems in Merging Phase Space, xv+331 pp. World Scientific, Singapore (2005)

Korolyuk, V., Swishchuk, A.: Semi-Markov Random Evolutions. Mathematics and Its Applications, vol. 308, x+310 pp. Kluwer, Dordrecht (1995) (English revised edition of Semi-Markov Random Evolutions, 254 pp. Naukova Dumka, Kiev (1992))

Korolyuk, V.S., Turbin, A.F.: On the asymptotic behaviour of the occupation time of a semi-Markov process in a reducible subset of states. Teor. Veroyatn. Mat. Stat. **2**, 133–143 (1970) (English translation in Theory Probab. Math. Stat. **2**, 133–143)

Korolyuk, V.S., Turbin, A.F.: Semi-Markov Processes and its Applications, 184 pp. Naukova Dumka, Kiev (1976)

Korolyuk, V.S., Turbin, A.F.: Mathematical Foundations of the State Lumping of Large Systems, 218 pp. Naukova Dumka, Kiev (1978) (English edition: Mathematical Foundations of the State Lumping of Large Systems. Mathematics and its Applications, vol. 264, x+278 pp. Kluwer, Dordrecht (1993))

Korolyuk, V.S., Brodi, S.M., Turbin, A.F.: Semi-Markov processes and their application. Probability Theory. Mathematical Statistics. Theoretical Cybernetics, vol. 11, pp. 47–97. VINTI, Moscow (1974)

Korolyuk, V.S., Penev, I.P., Turbin, A.F.: Asymptotic expansion for the distribution of the absorption time of a weakly inhomogeneous Markov chain. In: Korolyuk, V.S. (ed.) Analytic Methods of Investigation in Probability Theory, pp. 97–105. Akad. Nauk Ukr. SSR, Inst. Mat., Kiev (1981)

Kovalenko, I.N.: Studies in the Reliability Analysis of Complex Systems, 210 pp. Naukova Dumka, Kiev (1975)

Kovalenko, I.N.: Rare events in queuing theory – a survey. Queuing Syst. Theory Appl. **16**(1–2), 1–49 (1994)

Kovalenko, I.N., Kuznetsov, N.Yu., Shurenkov, V.M.: Models of Random Processes. A Handbook for Mathematicians and Engineers, 446 pp. CRC, Boca Raton, FL (1996) (A revised edition of the 1983 Russian original)

Kovalenko, I.N., Kuznetsov, N.Yu., Pegg, P.A.: Mathematical Theory of Reliability of Time Dependent Systems with Practical Applications. Wiley Series in Probability and Statistics, 316 pp. Wiley, New York (1997)

Lande, R., Engen, S., Saether, B-E.: Stochastic Population Dynamics in Ecology and Conservation. Oxford Series and Ecology and Evolution, x+212 pp. Oxford University Press, Oxford (2003)

Langville, A.N., Meyer, C.D.: Google's PageRank and Beyond: The Science of Search Engine Rankings, x+224 pp. Princeton University Press, Princeton, NJ (2006)

Latouche, G., Louchard, G.: Return times in nearly decomposable stochastic processes. J. Appl. Probab. **15**, 251–267 (1978)

Lévy, P.: Processus semi-Markoviens. In: Erven, P., Noordhoff, N.V. (eds.) Proceedings of the International Congress of Mathematicians, Vol. III. Amsterdam, 1954, pp. 416–426. North-Holland, Amsterdam (1954/1956)

Limnios, N., Oprişan, G.: Semi-Markov Processes and Reliability. Statistics for Industry and Technology, xii+222 pp. Birkhäuser, Boston (2001)

Marek, I., Pultarová, I.: A note on local and global convergence analysis of iterative aggregation-disaggregation methods. Linear Algebra Appl., **413**(2–3), 327–341 (2006)

Marek, I., Mayer, P., Pultarová, I.: Convergence issues in the theory and practice of iterative aggregation/disaggregation methods. Electron. Trans. Numer. Anal. **35**, 185–200 (2009)

Markushevich, A.I.: Theory of Functions of a Complex Variable. Vol. I, II, III. 2nd edn., xxiii+1138 pp. Chelsea, New York (1977, 1985) (English edition of Theory of Analytic Functions. GITTL, Moscow, 1950)

Meshalkin, L.D.: Limit theorems for Markov chains with a finite number of states. Teor. Veroyatn. Primen. **3**, 361–385 (1958) (English translation in Theory Probab. Appl. **3**, 335–357)

Meyer, C.D.: Stochastic complementation, uncoupling Markov chains, and the theory of nearly reducible systems. SIAM Rev. **31**(2), 240–272 (1989)

Meyn, S.P., Tweedie, R.L.: Markov Chains and Stochastic Stability, xxviii+594 pp. Cambridge University Press, Cambridge (2009) (2nd edition of Markov Chains and Stochastic Stability. Communications and Control Engineering Series, xvi+ 548 pp. Springer, London (1993))

Miller, B.L., Veinott Jr., A.F.: Discrete dynamic programming with a small interest rate. Ann. Math. Stat. **40**, 366–370 (1969)

Nåsell, I.: Extinction and Quasi-Stationarity in the Stochastic Logistic SIS Model. Lecture Notes in Mathematics, vol. 2022. Mathematical Biosciences Subseries, xii+199 pp. Springer, Heidelberg (2011)

Ni, Y.: Perturbed renewal equations with multivariate non-polynomial perturbations. In: Frenkel I., Gertsbakh I., Khvatskin L., Laslo Z., Lisnianski, A. (eds.) Proceedings of the International Symposium on Stochastic Models in Reliability Engineering, Life Science and Operations Management, Beer Sheva, pp. 754–763 (2010a)

Ni, Y.: Analytical and numerical studies of perturbed renewal equations with multivariate non-polynomial perturbations. J. Appl. Quant. Methods **5**(3), 498–515 (2010b)

Ni, Y.: Nonlinearly perturbed renewal equations: asymptotic results and applications. Doctoral Dissertation, 106, Mälardalen University, Västerås (2011)

Ni, Y.: Nonlinearly perturbed renewal equations: the non-polynomial case. Teor. Ĭmovirn. Mat. Stat. **84**, 111–122 (2012) (Also in Theory Probab. Math. Stat. **84**, 117–129)

Ni, Y.: Exponential asymptotical expansions for ruin probability in a classical risk process with non-polynomial perturbations. In: Silvestrov, D., Martin-Löf, A. (eds.) Modern Problems in Insurance Mathematics, chap. 6. EAA Series, pp. 67–91. Springer, Cham (2014)

Ni, Y., Silvestrov, D., Malyarenko, A.: Exponential asymptotics for nonlinearly perturbed renewal equation with non-polynomial perturbations. J. Numer. Appl. Math. **1**(96), 173–197 (2008)

Obzherin, Y.E., Boyko, E.G.: Semi-Markov Models. Control of Restorable Systems with Latent Failures, xiv+199 pp. Elsevier/Academic Press, Amsterdam (2015)

Pervozvanskiĭ, A.A., Smirnov, I.N.: An estimate of the steady state of a complex system with slowly varying constraints. Kibernetika (4), 45–51 (1974) (English translation in Cybernetics, **10**(4), 603–611)

Petersson, M.: Quasi-stationary distributions for perturbed discrete time regenerative processes. Teor. Ĭmovirn. Mat. Stat. **89**, 140–155 (2013) (Also in Theor. Probab. Math. Stat. **89**, 153–168)

Petersson, M.: Asymptotics of ruin probabilities for perturbed discrete time risk processes. In: Silvestrov, D., Martin-Löf, A. (eds.) Modern Problems in Insurance Mathematics, chap. 7, pp. 93–110. EAA Series, Springer, Cham (2014)

Petersson, M.: Asymptotic expansions for moment functionals of perturbed discrete time semi-Markov processes. In: Silvestrov, S., Rančić, M. (eds.) Engineering Mathematics II. Algebraic, Stochastic and Analysis Structures for Networks, Data Classification and Optimization, chap. 8, pp. 109–130. Springer Proceedings in Mathematics & Statistics, vol. 179. Springer, Cham (2016a)

Petersson, M.: Asymptotics for quasi-stationary distributions of perturbed discrete time semi-Markov processes. In: Silvestrov, S., Rančić, M. (eds.) Engineering Mathematics II. Algebraic and Analysis Structures for Networks, Data Classification, Optimization and Stochastic Processes, chap. 9. Springer Proceedings in Mathematics & Statistics, vol. 179, pp. 131–150. Springer, Cham (2016b)

Petersson, M.: Perturbed discrete time stochastic models. Doctoral Dissertation, Stockholm University (2016c)

Phillips, R.G., Kokotović, P.V.: A singular perturbation approach to modeling and control of Markov chains. IEEE Trans. Autom. Control **26**, 1087–1094 (1981)

Pinsky, M.A., Karlin, S.: An Introduction to Stochastic Modeling, 4th edn., x+563 pp. Elsevier/Academic Press, Amsterdam (2011)

Poliščuk, L.I., Turbin, A.F.: Asymptotic expansions for certain characteristics of semi-Markov processes. Teor. Veroyatn. Mat. Stat. **8**, 122–127 (1973) (English translation in Theory Probab. Math. Stat. **8**, 121–126)

Schweitzer, P.J.: Perturbation theory and finite Markov chains. J. Appl. Probab. **5**, 401–413 (1968)

Schweitzer, P.J.: Dual bounds on the equilibrium distribution of a finite Markov chain. J. Math. Anal. Appl. **126**(2), 478–482 (1987)

Schweitzer, P.J.: A survey of aggregation-disaggregation in large Markov chains. In: Stewart, W.J. (ed.) Numerical Solution of Markov Chains. Probability: Pure and Applied, vol. 8, pp. 63–88. Dekker, New York (1991)

Schweitzer, P.J., Kindle, K.W.: Iterative aggregation for solving undiscounted semi-Markovian reward processes. Commun. Stat. Stoch. Models **2**(1), 1–41 (1986)

Schweitzer, P., Stewart, G.W.: The Laurent expansion of pencils that are singular at the origin. Linear Algebra Appl. **183**, 237–254 (1993)

Schweitzer, P.J., Puterman, M.L., Kindle, K.W.: Iterative aggregation-disaggregation procedures for discounted semi-Markov reward processes. Oper. Res. **33**(3), 589–605 (1985)

Seneta, E.: Finite approximations to infinite non-negative matrices. Proc. Camb. Philol. Soc. **63**, 983–992 (1967)

Seneta, E.: Finite approximations to infinite non-negative matrices. II. Refinements and applications. Proc. Camb. Philol. Soc. **64**, 465–470 (1968a)

Seneta, E.: The principle of truncations in applied probability. Comment. Math. Univ. Carolinae **9**, 237–242 (1968b)

Seneta, E.: Nonnegative Matrices. An Introduction to Theory and Applications, x+214 pp. Wiley, New York (1973)

Seneta, E.: Iterative aggregation: convergence rate. Econ. Lett. **14**(4), 357–361 (1984)

Seneta, E.: Nonnegative Matrices and Markov Chains. Springer Series in Statistics, xvi+287 pp. Springer, New York (2006) (A revised reprint of 2nd edition of Nonnegative Matrices and Markov Chains. Springer Series in Statistics, xiii+279 pp. Springer, New York (1981))

Shurenkov, V.M.: Ergodic Markov Processes. Probability Theory and Mathematical Statistics, 332 pp. Nauka, Moscow (1989)

Silvestrov, D.S.: Limit Theorems for Composite Random Functions, 318 pp. Vysshaya Shkola and Izdatel'stvo Kievskogo Universiteta, Kiev (1974)

Silvestrov, D.S.: Semi-Markov Processes with a Discrete State Space. Library for the Engineer in Reliability, 272 pp. Sovetskoe Radio, Moscow (1980)

Silvestrov, D.S.: Exponential asymptotic for perturbed renewal equations. Teor. Ĭmovirn. Mat. Stat. **52**, 143–153 (1995) (English translation in Theory Probab. Math. Stat. **52**, 153–162)

Silvestrov D.S.: Limit Theorems for Randomly Stopped Stochastic Processes. Probability and Its Applications, xvi+398 pp. Springer, London (2004)

Silvestrov, D.S.: Asymptotic expansions for quasi-stationary distributions of nonlinearly perturbed semi-Markov processes. Theory Stoch. Process. **13**(1–2), 267–271 (2007)

Silvestrov D.S.: Nonlinearly perturbed stochastic processes and systems. In: Rykov, V., Balakrishnan, N., Nikulin, M. (eds.) Mathematical and Statistical Models and Methods in Reliability, chap. 2, pp. 19–38. Birkhäuser, Basel (2010)

Silvestrov, D.S., Abadov, Z.A.: Asymptotic behaviour for exponential moments of sums of random variables defined on exponentially ergodic Markov chains. Dokl. Acad. Nauk Ukr. SSR Ser. A (4), 23–25 (1984)

Silvestrov, D.S., Abadov, Z.A.: Uniform representations of exponential moments of sums of random variables defined on a Markov chain and of distributions of the times of attainment. I. Teor. Veroyatn. Mat. Stat. **45**, 108–127 (1991) (English translation in Theory Probab. Math. Stat., **45**, 105–120)

Silvestrov, D.S., Abadov, Z.A.: Uniform representations of exponential moments of sums of random variables defined on a Markov chain and of distributions of the times of attainment. II. Teor. Veroyatn. Mat. Stat. **48**, 175–183 (1993) (English translation in Theory Probab. Math. Stat. **48**, 125–130)

Silvestrov, D., Manca, R.: Reward algorithms for semi-Markov processes. Method. Comp. Appl. Probab. DOI 10.1007/s11009-017-9559-2, 19 pp. (2017)

Silvestrov, D.S., Petersson, M.: Exponential expansions for perturbed discrete time renewal equations. In: Karagrigoriou, A., Lisnianski, A., Kleyner, A., Frenkel, I. (eds.) Applied Reliability Engineering and Risk Analysis. Probabilistic Models and Statistical Inference, chap. 23, pp. 349–362. Wiley, New York (2013)

Silvestrov, D., Silvestrov, S.: Asymptotic expansions for stationary distributions of perturbed semi-Markov processes. In: Silvestrov, S., Rančić, M. (eds.) Engineering Mathematics II. Algebraic, Stochastic and Analysis Structures for Networks, Data Classification and Optimization, chap. 10. Springer Proceedings in Mathematics & Statistics, vol. 179, pp. 151–222. Springer, Cham (2016a)

Silvestrov, D., Silvestrov, S.: Asymptotic expansions for stationary distributions of nonlinearly perturbed semi-Markov processes. I, II. Part I: arXiv:1603.04734, 30 pp., Part II: arXiv:1603.04743, 33 pp. (2016b)

Silvestrov, D.S., Petersson, M., Hössjer, O.: Nonlinearly perturbed birth-death-type models. Research Report 2016:6. Department of Mathematics, Stockholm University, 63 pp. (2016). arXiv:1604.02295

Simon, H.A., Ando, A.: Aggregation of variables in dynamic systems. Econometrica **29**, 111–138 (1961)

Smith, W.L.: Regenerative stochastic processes. Proc. R. Soc. Lond. Ser. A **232**, 6–31 (1955)

Stewart, G.W.: Gaussian elimination, perturbation theory, and Markov chains. In: Meyer, C.D., Plemmons, R.J. (eds.) Linear Algebra, Markov Chains, and Queueing Models. IMA Volumes in Mathematics and its Applications, vol. 48, pp. 59–69. Springer, New York (1993a)

Stewart, G.W.: On the perturbation of Markov chains with nearly transient states. Numer. Math. **65**(1), 135–141 (1993b)

Stewart, G.W.: Matrix Algorithms. Vol. I. Basic Decompositions, xx+458 pp. SIAM, Philadelphia, PA (1998)

Stewart, G.W.: Matrix Algorithms. Vol. II. Eigensystems, xx+469 pp. SIAM, Philadelphia, PA (2001)

Stewart, G.W., Sun, J.G.: Matrix Perturbation Theory. Computer Science and Scientific Computing, xvi+365 pp. Academic Press, Boston (1990)

Stewart, G.W., Zhang, G.: On a direct method for the solution of nearly uncoupled Markov chains. Numer. Math. **59**(1), 1–11 (1991)

Sumita, U., Reiders, M.: A new algorithm for computing the ergodic probability vector for large Markov chains: replacement process approach. Probab. Eng. Inf. Sci. **4**, 89–116 (1988)

Takács, L.: Some investigations concerning recurrent stochastic processes of a certain type. Magyar. Tud. Akad. Mat. Kutató Int. Kŏzl. **3**, 114–128 (1954)

Turbin, A.F.: An application of the theory of perturbations of linear operators to the solution of certain problems that are connected with Markov chains and semi-Markov processes. Teor. Veroyatn. Mat. Stat., **6**, 118–128 (1972) (English translation in Theory Probab. Math. Stat. **6**, 119–130)

Van Doorn, E.A.: Stochastic Monotonicity and Queueing Applications of Birth-Death Processes. Lecture Notes in Statistics, vol. 4, vi+118 pp. Springer, New York (1981)

Vantilborgh, H.: Aggregation with an error of $O(\varepsilon^2)$. J. Assoc. Comput. Mach. **32**(1), 162–190 (1985)

Veinott Jr., A.F.: Discrete dynamic programming with sensitive discount optimality criteria. Ann. Math. Stat. **40**, 1635–1660 (1969)

Volker, N.: Semi-Markov Processes. Scientific Paperbacks, Mathematics/Physics Series, vol. 260, 162 pp. Akademie-Verlag, Berlin (1980)

Whitt, W.: Stochastic-Process Limits. An Introduction to Stochastic-Process Limits and Their Application to Queues. Springer Series in Operations Research, xxii+601 pp. Springer, New York (2002)

Yin, G., Zhang, Q.: Discrete-time singularly perturbed Markov chains. In: Yao, D.D., Zhang, H., Zhou, X.Y. (eds.) Stochastic Modeling and Optimization, pp. 1–42. Springer, New York (2003)

Yin, G.G., Zhang, Q.: Discrete-Time Markov Chains. Two-Time-Scale Methods and Applications. Stochastic Modelling and Applied Probability, xix+348 pp. Springer, New York (2005)

Yin, G.G., Zhang, Q.: Continuous-Time Markov Chains and Applications. A Two-Time-Scale Approach. Stochastic Modelling and Applied Probability, vol. 37, xxii+427 pp. Springer, New York (2013) (2nd revised edition of Continuous-Time Markov Chains and Applications. A Singular Perturbation Approach. Applications of Mathematics, vol. 37, xvi+349 pp. Springer, New York (1998))

Yin, G., Zhang, Q., Yang, H., Yin, K.: Discrete-time dynamic systems arising from singularly perturbed Markov chains. In: Proceedings of the Third World Congress of Nonlinear Analysts, Part 7, Catania, 2000. Nonlinear Analysis, vol. 47(7), pp. 4763–4774 (2001)

Index

D. Silvestrov, S. Silvestrov, *Nonlinearly Perturbed Semi-Markov Processes*,
SpringerBriefs in Probability and Mathematical Statistics,
DOI 10.1007/978-3-319-60988-1

Printed in the United States
By Bookmasters